T0318547

Materials Experience 2

Materials Experience 2

Expanding Territories of Materials and Design

Edited by

Owain Pedgley

Department of Industrial Design, Middle East Technical University, Ankara, Turkey

Valentina Rognoli

Design Department, Politecnico di Milano, Milan, Italy

Elvin Karana

Faculty of Industrial Design Engineering, Delft University of Technology, Delft, Netherlands

Butterworth-Heinemann is an imprint of Elsevier
The Boulevard, Langford Lane, Kidlington, Oxford OX5 1GB, United Kingdom
50 Hampshire Street, 5th Floor, Cambridge, MA 02139, United States

British Library Cataloguing-in-Publication Data
A catalogue record for this book is available from the British Library

Library of Congress Cataloging-in-Publication Data
A catalog record for this book is available from the Library of Congress

ISBN: 978-0-12-819244-3

For Information on all Butterworth-Heinemann publications
visit our website at https://www.elsevier.com/books-and-journals

Publisher: Matthew Deans
Acquisitions Editor: Christina Gifford
Editorial Project Manager: Marina L. Kuhl
Production Project Manager: Sojan P. Pazhayattil
Cover Designer: Matthew Limbert

Typeset by MPS Limited, Chennai, India

Contents

Around The Corner: Recent and Ongoing Projects in Materials and Design

List of Contributors

Ozgun Kilic Afsar
Media Technology and Interaction Design, KTH Royal Institute of Technology, Stockholm, Sweden

Bilge Merve Aktaș
Department of Design, Aalto University, Espoo, Finland

Camilo Ayala-Garcia
Department of Design, Universidad de los Andes, Bogotá, Colombia

Conny Bakker
Faculty of Industrial Design, University of Technology, Delft, Netherlands

Ruud Balkenende
Faculty of Industrial Design, University of Technology, Delft, Netherlands

Bahareh Barati
Faculty of Industrial Design Engineering, Delft University of Technology, Delft, Netherlands

Ben Bridgens
School of Architecture, Planning and Landscape, Newcastle University, Newcastle upon Tyne, United Kingdom

Blaine Brownell
University of North Carolina at Charlotte, Charlotte, United States of America

Carole Collet
Living Systems Lab, Central Saint Martins UAL, London, United Kingdom

Camilla Groth
Department of Visual and Performing Arts Education, University of South-Eastern Norway, Notodden, Norway

Markus Holzbach
IMD Institute for Materialdesign University of Art and Design Offenbach, Offenbach am Main, Germany

SunMin May Hwang
Human Factors and Ergonomics, University of Minnesota, Minneapolis, United States of America

Alexandra Karakas
Doctoral School of Philosophy, Eotvos Lorand University, Budapest, Hungary

Elvin Karana
Faculty of Industrial Design Engineering, Delft University of Technology, Delft, Netherlands

Michelle Knox
Department of Medicine, University of Alberta, Edmonton, AB, Canada

Manuel Kretzer
Materiability Research Group, Dessau Department of Design Anhalt, University of Applied Sciences, Dessau, Germany

Carla Langella
Department of Architecture and Industrial Design (DADI), University of Campania "Luigi Vanvitelli", Caserta, Italy

Debra Lilley
School of Design and Creative Arts, Loughborough University, Loughborough, United Kingdom

Vincenzo Maselli
Department of Planning, Design and Technology of Architecture, Sapienza University of Rome, Rome, Italy

Jussi Ville Mikkonen
Department of Electronics and Communications Engineering, Tampere University of Technology, Tampere, Finland

Nithikul Nimkulrat
OCAD University, Toronto, Canada

Mutian Niu
School of Film and TV Arts, Xi'an Jiaotong-Liverpool University, Suzhou, P.R. China

Francesca Ostuzzi
Department of Industrial Systems Engineering and Product Design, Ghent University (Campus Kortrijk), Ghent, Belgium

Dilan Ozkan
School of Architecture, Planning and Landscape, Newcastle University, Newcastle upon Tyne, United Kingdom

Doriana Dal Palù
Department of Architecture and Design, Politecnico di Torino, Torino, Italy

Stefano Parisi
Design Department, Politecnico di Milano, Milan, Italy

Owain Pedgley
Department of Industrial Design, Middle East Technical University, Ankara, Turkey

Barbara Pollini
Design Department, Politecnico di Milano, Milan, Italy

Emmi Anna Maria Pouta
Department of Design/Department of Communications and Networking, Aalto University, Espoo, Finland

Dilusha Rajapakse
School of Arts and Design, Nottingham Trent University, Nottingham, United Kingdom

Valentina Rognoli
Design Department, Politecnico di Milano, Milan, Italy

Indji Selim
Department of Industrial Design, Ss. Cyril and Methodious University Skopje, Skopje, Republic of North Macedonia

Bahar Șener
Department of Industrial Design, Middle East Technical University, Ankara, Turkey

Marie Louise Juul Søndergaard
Media Technology and Interaction Design, KTH Royal Institute of Technology, Stockholm, Sweden

Assia Stefanova
School of Architecture, Planning and Landscape, Newcastle University, Newcastle upon Tyne, United Kingdom

Ziyu Zhou
Design Department, Politecnico di Milano, Milan, Italy

Biographies

Owain Pedgley is Professor of Industrial Design at Middle East Technical University, Ankara. His expertise is in design for interaction and user experiences, with specialization in two areas of application: materials and materialization, and musical instrument design and innovation. He has published widely in these areas and coordinates projects as a researcher, thesis supervisor, and instructor. Owain is a strong advocate of research through design (RtD), having championed this approach to academic studies for over 20 years through publications and supervisions, traced back to early adoption of RtD in his own PhD (1999). He is a member of the Editorial Board of *Design Studies* and an Associate Editor of *She Ji: The Journal of Design, Economics and Innovation*. He is also an Honorary Research Fellow in the School of Engineering at the University of Liverpool, having served as a founding member of its Industrial Design BEng/MEng programs (2014–17). Prior to commencing his academic career, Owain worked as a product designer in the sports equipment and musical instrument sectors, co-founding the polymer acoustic guitar innovation venture "Cool Acoustics."

(*Image Credit: Bahar Şener-Pedgley*)

Valentina Rognoli is Associate Professor in the Design Department at Politecnico di Milano. She is a pioneer in the field of materials experience, starting almost 20 years ago and has established internationally recognized expertise on the topic in both research and education. Her mission is raising sensibility and making professional designers and future designers conscious of the infinite potential of materials and processes. The investigations of her research group focus on pioneering and challenging topics including DIY-Materials for social innovation and sustainability; bio and circular materials; urban materials and materials from waste and food waste; materials for interactions and IoT (ICS Materials); speculative materials; tinkering with materials; materials-driven design method; CMF design; emerging materials' experiences; and material education in the field of design. Since 2015, Valentina jointly leads, with Elvin Karana, the international research group *Materials Experience Lab*. She participated as a principal investigator in the European Project *Made*, co-funded by the Creative Europe Program of The European Union, which aimed to boost talents toward circular economies across Europe. Valentina is the author of over 50 publications. She has organized international workshops and events and has contributed as an invited speaker and reviewer for relevant journals and international conferences.

(*Image Credit: Lab Image Design Polimi*)

Elvin Karana is Professor of Materials Innovation and Design in the Faculty of Industrial Design Engineering at Delft University of Technology. Giving emphasis to materials' role in design as experiential and yet deeply rooted in their inherent properties, Elvin explores and navigates the productive shifts between materials science and design for materials and product development in synergy. In 2015, she founded the cross-country research group *Materials Experience Lab*, which she leads jointly with Valentina Rognoli. Elvin has over 70 scientific publications in peer-reviewed journals and conferences. Her recent book *Still Alive* (2020) brings to the attention a new and exciting design space, where she proposes that by discovering ways to maintain an organism's aliveness as a tangible manifestation of a biodesign process, *livingness* will become a persistent material quality in design. In 2019, she founded the biodesign research lab *Material Incubator* ([MI] Lab) that invites designers to harness the potential of living organisms for unique functionalities, interactions, and expressions in the everyday. Material Incubator brings together researchers and practitioners from Avans University of Applied Sciences and Delft University of Technology.

(*Image Credit: Elvin Karana*)

Foreword

I am writing this Foreword in the middle of the COVID-19 global pandemic. My family and I are locked down in our flat in London. The schools are shut, and I spend my time swapping between helping my kids interact with their teachers online, to interacting with my own university students, again through a screen. Over many months of doing this, I have realized that although online teaching has been a lifeline, it is no way a substitute for real face-to-face teaching. This is not because the medium is deficient for communicating ideas and information. What gets lost is meaning. This seems odd because other screen-based activities such as TV programs and films are able to communicate meaning very well. But somehow an online lecture, or workshop, seems to be a simulation of real meaningful interaction. The physical classroom, with its sounds, smells, and sensations, all contribute to something very different—it is a material experience. For this and many other reasons, most people accept the experience of face-to-face education is fundamentally different from online learning and think no more. But designers and materials researchers find it impossible to resist exploring the origin of such differences. If you are one of these people, then this book and its predecessor, *Materials Experience*, are for you.

That the COVID-19 coronavirus is transmitted through material interaction, either through touch or through the air we all breathe, is the reason why schools are closed. The virus itself is invisible but its presence or potential presence has made us hyper-aware of the physicality of interacting with each other in schools, offices, or shops. Clasping the handle of a door now has an element of jeopardy that it lacked before. On the way into a cafe, we wonder how many infected people touched the door handle before us. We know the risk of infection is real not just because other people have told us that small droplets of liquid can pass between people that touch the same things. We know it is real because of the millions of times we ourselves have touched a surface and left a smudge on it. I have to clean my spectacles several times a day for this reason, and the screen of my phone too. Our materials experience of surfaces and their other affordances such as their ability to attract, extract, and collect oils and essences from our hands (viruses and all) is learnt this way. It is this experience that gives credence to the danger of viruses being transmitted via touch. And yet this same touch is also how we transmit our affection and feelings for each other. They literally go hand-in-hand, communication, and contamination. And so the pandemic has not only closed schools but emotionally sealed us off from each other. It is one of the many tragedies of the COVID-19 pandemic that millions have died in hospital without anyone hugging them goodbye as they slipped out of this world.

When we reach in our bag for a small bottle of cool alcohol gel and squirt it onto our hands, we experience a feeling of protection that is emphasized by the coolness of gel. This is due to the evaporation of the alcohol on our hands, removing energy from our hands through the flow of latent heat. To design a materials experience like that needs not only the scientific knowledge of how to make a liquid that will destroy the virus and then vanish, but also an understanding of how humans react to coolness, stickiness, and slipperiness. The pandemic has been a lesson for me on the good, the bad, and the ugly of alcohol sanitizers. Some are too runny and gush out leaving me soaked and annoyed. Others stick to my hands but then do not evaporate fast enough and leave my hands wet for an annoying amount of time. Others evaporate fast but leave my hands sticky and uncomfortable. I have had so many bad experiences that when I find a hand sanitizer that cleans, cools, and then disappears in seconds, I feel the need to shake the hand of the person who designed such a great material experience. That person, or more likely team of people, understands the complexity of material experience and how to design products using that knowledge. It is one of my hopes that more people will read this book and understand the mix of art, science, engineering, and cultural understanding required to design better products.

The daily wearing of a mask is another materials experience that is new to me with the pandemic. Through experimentation I have found the combination of materials and mask design that suits me, my beard, and my glasses that steam up. But the biggest discovery for me has been the relaxing nature of mask-wearing. No longer do I feel the need to smile out of politeness at every public interaction. Nor do I feel that my irritation with a particular situation will be written clearly on my face. To be muted like this does not feel like oppression. It is restful, serene even. It is a freedom, a privacy, and a protection. No doubt I would feel different if my mask-wearing became permanent, which it is of course to many in the world. The materials experience of mask-wearing has made me much more interested in the cultural wearing of masks, as well as those in the medical profession for whom it is part of a daily uniform worn 8–10 hours a day. It is my hope too that the designers of the masks of the future will read this book.

My purpose for recounting my impressions of lockdown life in a pandemic is to emphasize first that our lives are always changing, and this brings us new materials experiences. But also, to show that the role of product designers and materials researchers is so fundamental. We need communities of researchers well versed in understanding materials experiences, spanning their cultural, social, scientific, and engineered aspects. The physical world is an almost infinite realm of experience and we need people able to navigate it to design the products and materials of the future. This necessitates understanding different disciplinary approaches to materials properties and making, as well as the interfaces between them and the living world. There is no better guide out there to this approach than the *Materials Experience* books, and I am very happy to see this second book continuing to provide a guide for future designers and materials researchers.

Mark Miodownik,
Director, UCL Institute of Making
February 2021

Preface

Seven years have passed since the publication of the first volume of *Materials Experience*, subtitled *Fundamentals of Materials and Design*. We had in mind that the book would bring together a wide range of views from an equally wide range of contributors—spanning design, engineering, architecture, user experience, materials science, and education—all focused on opening a discussion on what the "experiential side" of materials is and what it involves. Gathering the views and experiences of practicing designers as well as scholars was a major motivation for the book. As a result, the book struck a chord among academic and professional design communities for whom materials are much more than a substance that serves to give physical form to an idea.

We started out with no timetable to put together a successor to *Materials Experience*. Our view was that it needed to be "out there" for some time before we could properly gauge its influence and understand which of the various concepts and streams of work embedded within would take root and deliver new research. As it happens, all Editors converged on the idea in 2018 that a new book was merited. We felt that a substantial body of knowledge had developed in the intervening years and that a new book was warranted to catch it into a single location. Serendipitously, we were invited by Elsevier to prepare a follow-up book. So, spurred on through a combination of circumstances, we put together plans for this companion volume to *Materials Experience*—envisioned as a new set of original works—which would continue the lineage started with the first volume.

Materials Experience 2: Expanding Territories of Materials and Design is organized as a "sandwich" of three sections. The first and last sections comprise peer-reviewed chapters from invited authors on a variety of contemporary topics in materials experience, especially those that we identified as emergent or significantly developed since the first volume of the book.

The middle section, titled "Around the Corner," presents a shortlist of peer-reviewed projects following an open call to doctoral students and early career researchers to disseminate their work through the book. We are especially pleased with the results: an energetic collection spanning current or recently completed research in materials and design, collectively providing an international view not only of contemporary subjects but also hints for where subjects will emerge or expand in the future. The project contributors represent institutions from around the globe: Belgium, Canada, China, Finland, Hungary, Italy, the Netherlands, North Macedonia, Sweden, the United Kingdom, and the United States.

As academics, we grow through nurturing discussions and interactions with our colleagues and our students. We would like to thank our institutions for the opportunity to develop intellectually through our research and educational activities: the Department of Industrial Design at Middle East Technical University, Turkey; the Department of Design at Politecnico di Milano, Italy; and the Department of Sustainable Design Engineering at Delft University of Technology, the Netherlands. We would like to thank all contributors to this volume of *Materials Experience* for their support, dedication, and patience during the process—especially when having to work under the exceptional demands that the COVID-19 pandemic has brought. We would also like to highlight special gratitude to Mark Miodownik for picking through the book draft to prepare his Foreword.

Our thanks extend to the core team at Elsevier, to whom we often made special requests, which were gracefully granted: Christina Gifford, Mariana Kühl Leme, and Indhumathi Mani. We thank you for your patience, and for having to deal with the pickiness of "designer" Editors.

As always, our final words of thanks must be reserved for our immediate family members who have encouraged and tolerated us during the process of turning the idea of a second volume into reality: Bahar Şener-Pedgley (alongside Jessica and Lucas), Yunier Virelles (alongside Ernesto and Camilo), and Jaap Rutten (alongside Elis and Eren).

Editors
January 2021

CHAPTER 1

Expanding territories of materials and design

Owain Pedgley[a], Valentina Rognoli[b], Elvin Karana[c]
[a]Department of Industrial Design, Middle East Technical University, Ankara, Turkey,
[b]Design Department, Politecnico di Milano, Milan, Italy,
[c]Faculty of Industrial Design Engineering, Delft University of Technology, Delft, Netherlands

1.1 Introduction

People's experience of materials, regarding their here-and-now and possible futures, is largely bound into complex accounts of how materials are mobilized in the design of artifacts. The first volume of *Materials Experience*, subtitled *Fundamentals of Materials and Design* (Karana et al., 2014), focused on describing people–material relationships, with the central premise that materials experience can be viewed from the perspective of the designer who creates artifacts and from the perspective of people who own and interact with those artifacts.

In this second volume, we have drawn upon our observations of how materials experience as a concept has evolved and been mobilized to incorporate new ways of thinking and doing in design. We have subtitled the book *Expanding Territories of Materials and Design*, encompassing a critical perspective on the changing role of design/designers, the increased prevalence of material-driven design (MDD) practices, and the increasing attention among design scholars to the role of materials themselves as active and influential agents within and outside design processes. *Materials Experience 2* is therefore a companion to the first volume. In this introductory chapter, to benefit readers venturing into the field of materials experience, we first provide a concise account of where materials experience originated from, alongside its main concepts.

1.1.1 Lineage of materials experience

Materials experience as a topic of discussion can be traced to two landmark books, published in different decades: Manzini (1986) and Ashby & Johnson (2002); Manzini (1986) was the first author, and for a long time also the only author, to present convincing arguments and examples on the interrelationships between materials, inventions, and people's experiences. His work

Materials Experience 2. DOI: https://doi.org/10.1016/B978-0-12-819244-3.00028-4

emphasizes the role of materials, which, when combined with design, can invent new solutions and meaningful experiences. The book stands as a milestone in this area of research. It confirmed the importance of materials in the design process, in addition to their autonomy: that is, the fact that materials can be considered design objects in themselves.

In 2002 Ashby and Johnson provided the first textbook on materials and design that purposefully adopted a "designerly" view of the materials world, in which product esthetics were discussed alongside technical possibilities and engineering material properties. Their work was particularly significant because of its heritage and interdisciplinarity—born from Ashby's world-renowned expertise in the Engineering Department of Cambridge University and Granta Design Ltd—but reaching out into product design through Johnson's doctoral studies and connections with the innovation consultancy IDEO. This book, which has been revised through multiple editions, remains crucial because it highlights very strongly how materials and design benefit from interdisciplinary approaches.

Around the time of Ashby and Johnson's publication, design research on the subjects of user experience, materials selection, materials information, and material sensory qualities started to expand significantly. In 2004 Rognoli's doctoral thesis (Rognoli, 2004) contributed to this field providing a vocabulary and a tool for designers to understand the sensorial qualities of materials. Karana et al. (2008) coined the term "materials experience" to describe the experiences that people have with, and through, the materials of a product. Subsequently, in her thesis, entitled *Meanings of Materials*, Karana (2009) conducted a series of studies providing empirical evidence on the multifaceted system that underpins our material experiences. Materials experience is therefore firmly rooted within the practices and research activities of product design. Its fundamentals formed through a fusion of the aforementioned materials and design studies. The subsequent years saw a sharp increase in research and general interest in materials experience, across a variety of design specialties and professions, beyond its origins in product design. The aim of *Materials Experience: Fundamentals of Materials and Design* (Karana et al., 2014) was to bring together these various scientific developments and practical insights into a single location, to construct a comprehensive view on the concept of materials experience.

Readers are recommended to refer to earlier publications for a lengthier discussion on the drivers and origins of materials experience: to the article *On Materials Experience* in the journal *Design Issues* (Karana et al., 2015a), which itself was a development of the *Introduction* in the original volume of *Materials Experience*; and to the *Editorial* of the special issue on *Emerging Material Experiences* in the journal *Materials and Design* (Karana et al., 2016), in which it was argued that designing with new materials through the lens of "materials experience" can be a powerful strategy to introduce those materials to societies. In the context of design education, a number of tools and techniques have been developed to make materials experience thinking more actionable in design thinking (Pedgley et al., 2016). Veelaert et al. (2020) published an extensive analysis of 64 experiential material characterization studies in the journal *Materials and Design*, which clearly indicates how materials experience research has become prominent among design scholars over the past few years. We anticipate the trend to continue, especially given the emergence of new kinds of materials into everyday experiences.

1.1.2 Materials experience framework

In its original description, materials experience consists of three experiential levels: esthetic (sensorial) experience (e.g., we find materials cold, smooth, and shiny), experience of meaning (e.g., we think materials are modern, sexy, and cozy), and emotional experience (e.g., materials make us feel amazed, surprised, and bored). We used these levels to ground the first volume of *Materials Experience*. In 2015 the performative level was added as a fourth level of materials experience, inviting scholars to explore how new and emerging materials in the everyday may elicit novel actions and, ultimately, practices (Giaccardi & Karana, 2015). Accordingly, the expanded four-level framework emphasizes that a comprehensive definition of materials experience should acknowledge the active role of materials not only in shaping our internal dialogs with artifacts, but also in shaping our ways of doing.

The addition of the performative level has been a springboard for many fruitful and diverse research studies. It has opened up possibilities to design for and with novel materials through an interactional lens. For example, one study has looked at the use of mycelium-based composites that offer a new practice for unpacking (wine) bottles (Fig. 1.1), by inviting people to pluck and pick the material to reveal its contents (Karana et al., 2018). The qualities of the material are fine-tuned to guide this response.

The performative level is also explicit in the domain of Interactive Connected Smart (ICS) Materials, which are hybrid material systems that exhibit adaptive and responsive qualities triggered by environmental inputs or through interaction with people (Ferrara et al., 2018). These materials imply and enable novel material experiences relying on dynamism, surprise, and, among others, performativity (Parisi & Rognoli, 2021; Rognoli, 2020). For example, one project developed under the umbrella of ICS Materials investigation has aimed to create do-it-yourself (DIY) conductive ink to understand its properties, opportunities, and limitations (Fig. 1.2).

FIGURE 1.1
An innovative packaging for (wine) bottles, designed with mycelium-based materials (Karana et al., 2018), by Davine Blauwhoff, 2017/18 Master's Graduation Project, Industrial Design Engineering, Delft University of Technology. *Reproduced with permission from Davine Blauwhoff.*

Metamaterials are another class of material for which the performative level of materials experience is highly relevant. These materials are designed with complex internal geometric microstructures and typically realized through high-resolution 3D printing (Ion & Baudisch, 2020; Ion et al., 2018). The materials deform and reshape in a controllable way when exposed to an input such as a force. Such dynamic qualities of the material can be mobilized to evoke certain actions from people, for example, through changes in surface texture (Fig. 1.3).

1.2 Expanding territories

If we scan across the contributions to this second volume of *Materials Experience*, some noticeable themes become evident that were either absent in the first volume or less prominent. We have defined the themes as the expanding territories of materials and design: the changing role of design and designers as well as the changing role of materials in design. Unsurprisingly, material implications in achieving sustainability goals continue to be a connective theme among many of the contributions, with a detectable sense of urgency in tone.

1.2.1 Changing role of design and designers

Material skills are (re)gaining importance in design. Hands-on material manipulation and evaluation has a long history with design. Crafts practices

FIGURE 1.2
L-INK project—"test, learn, hack" project by Gaia Bianco, 2017/18 Master's Graduation Project, Product Design for Innovation, Politecnico di Milano, supervised by Valentina Rognoli and Stefano Parisi. *Reproduced with permission from Gaia Bianco.*

are built on such relations, with craftspeople shaping materials into often unique artifacts. With the arrival of industrial design came a separation between intent (design) and making (production). The separation did not remove the designer's involvement with materials, but it did bring about changes. Propositional knowledge on material properties, material processing, and material applications became dominant because designers were no longer also the makers. Instead, designers were required to hand-over designs that were materially feasible and producible by someone else, somewhere else. In product design education, particularly at technical universities,

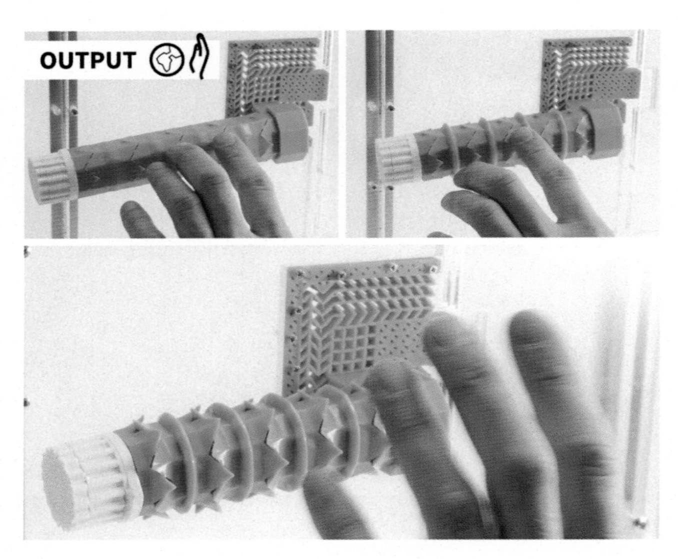

FIGURE 1.3
Door handle from metamaterial, resulting in dynamic surface texture changes and tactile communication. *Reproduced with permission from Alexandra Ion/Hasso Plattner Institute.*

it became normal to be *informed about materials*, rather than have *experience with or through materials*. It was (and still is) convenient to teach and learn in this manner since the requirements for workshops, technicians, and access to materials are minimized. In this respect, educational institutions in the art and design tradition have been much more resilient in maintaining material skills and direct material contact, placing them at the core of the student experience.

In recent years, the rise of maker spaces and hacker/repair hubs has contributed to a renewed emphasis on material skills and first-hand physical material experiences. Material libraries, today's ubiquitous and organized realities intended to be catalysts of reflection on design materials, have turned out to be insufficient for the needs of designers who have shown willingness not only to select a material but also to create and transform it directly. We now see the emergence of designers who are no longer satisfied with choosing or selecting materials, but now create materials independently, with a spirit of innovation in the language and the choice of resources and raw materials (see, e.g., the work of renowned designers Diana Scherer, Maurizo Montalti, and Natsai Audrey Chieza, among others).

From this standpoint, material explorations and material tinkering (Karana et al., 2015b; Parisi et al., 2017; Rognoli & Parisi, 2021) are talked about as no less important than materials selection. In the contemporary *material atelier*, designers and designers-in-training become acquainted with materials in highly practical ways. They tinker with the existing materials and engage in DIY material creation, namely, *DIY Materials* (Ayala-Garcia et al., 2017; Rognoli et al., 2015) (Fig. 1.4).

When the principles of personal materials tinkering and DIY material-making are scaled up to become professional practices in and of themselves, designers become the creators of new materials as much as creators of new products embodying those materials. In this journey, designers are increasingly cooperating with materials scientists, engineers, and biologists—traditionally separated communities—for the common purpose of designing new materials. It is not that design communities in the past have failed to engage with other disciplines to develop materials—there are successful examples (e.g., Franinović & Franzke, 2015; Miodownik & Tempelman, 2013). The difference in recent times is that these cooperations are giving rise to a new figure, the materials designer, with distinctive new forms of design practice.

In fact, the figure of the material designer had already been theorized in Italy as early as the 1980s and realized in the 1990s (Manzini & Petrillo, 1991; Trini Castelli & Petrillo, 1985; Manzini, 1996). The focus then was on material expressions, whereas today "materials design" is emerging as a new transdiscipline, with material designers as transdisciplinary practitioners who can design materials with unique functionalities and experiential qualities.

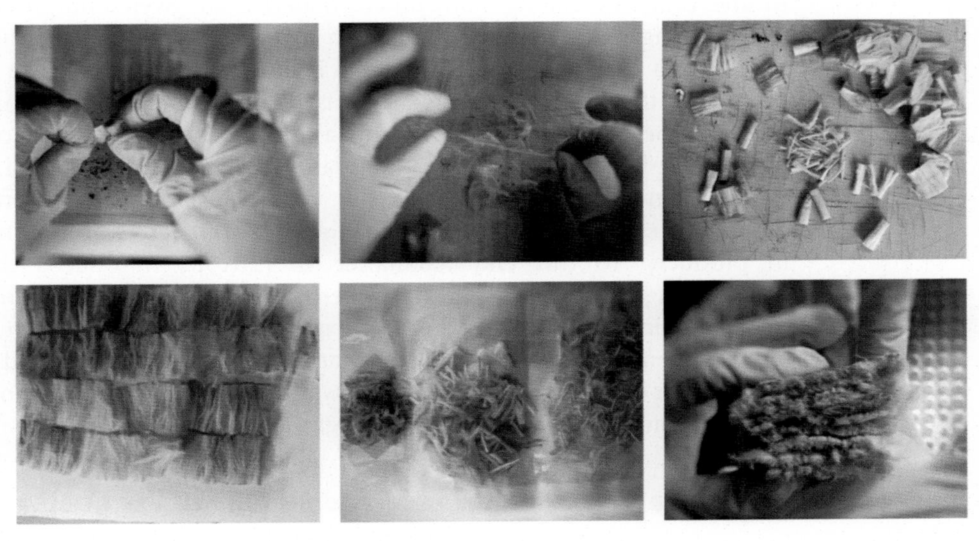

FIGURE 1.4

A DIY-Material example from cigarette butt materials. "Butts Bunny" by Carolina Giorgiani, Jinan Jezzini, Davide Mosito, Designing Materials Experience Course 2016/17, School of Design, Politecnico di Milano. *Reproduced with permission from Carolina Giorgiani.*

A major difference compared with the 1990s is that it is now far easier to take this path, thanks to the democratization of technologies, easier access to tools and above all access to information (courtesy of the Internet and social networks). Hence, boundaries between the "design of materials" and the "materials of design" fade, to become united under the common endeavor of exploring novel form, function and experience *potentials of materials* and their *process-abilities* (Barati & Karana, 2019).

1.2.2 Changing role of materials

Materials for design have traditionally been sourced and selected from among natural inert, once-living, or synthetic matter. In recent years, living materials have come to the fore of research and development. Emerging bio-design and biofabrication practices centered on growing materials (Camere & Karana, 2018; Ginsberg & Chieza, 2018; Collet, 2013; Myers, 2012), or facilitating the growth of *living artifacts* (Karana, 2020; Karana et al., 2020), present a shift in our understanding of what a material can be and the processes by which a product becomes materialized. In these practices, the designer assumes the role of co-creator of artifacts with living organisms. The designer still exerts intent, but also steps back to delegate responsibility to the material itself to grow into a design-relevant entity. A shift takes place: from being a materials designer to a facilitator of the grown, as if she or he was a breeder.

In such practices, among many others, materials are given a purposefully special standing: they can form the launch pad for ideation and be the "raison d'être" for a project. When the material is the star, designers engage in a process that is the antithesis of a reductionist, selection-based approach to materials. In other words materials become a fruitful starting point for conceptual product design. However, the utilitarian and expressive potential of new, unusual, and emerging materials can be challenging to envision and to design for, especially when compared to familiar materials commonly found in products. To this end Karana et al. (2015b) introduced the *MDD Method*—suggesting techniques to understand and integrate both the technical properties of a material and its experiential qualities in unveiling the design potential of the material. Motivating a return to making and material tinkering in design, the MDD Method was designed with the aim to give the designer guidelines for creating meaningful materials experiences.

1.2.3 Materials experience and sustainability goals

Environmental problems are increasingly evident in the world. Societal interest and comprehension of sustainability are probably higher than ever. The responsibility of designers is to transition toward sustainability goals by proposing solutions that include environmental reflection on materials. The role of materials experience in the context of sustainability has already been investigated using the concept of "Materially Yours" (Karana et al., 2017), in which strategies are indicated to suggest ways of using materials at different experiential levels to assist in the design of longer-lasting products. By expanding the territories of materials experience, we hope that the ability for designers to address environmental considerations will be widened. The growing impact of our objects (and waste) and the ever-increasing pressure on material resources requires a change in our relationship with the objects themselves and the materials they are made of.

The circular economy, low carbon footprints, reduced consumption, and local sourcing are terms and concepts that have been clearly explained and promoted for some considerable time as ways to reach sustainability goals and keep material usage in check. Yet, the responsibility and accountability of the individual designer has been difficult to tie down. What is clear is that there are currently renewed efforts, initiatives, and impetus to deliver impactful change in this regard, against a backdrop of severe consequences of inaction. Sustainability goals are integral to many of the contributions to this volume of *Materials Experience*. For example, through the lens of materials experience, a thorough understanding of people's acceptance of new and alternative materials can be gained, as well as people's willingness to engage as product and material (resource) custodians rather than owners of products or consumers of material (Karana, 2020).

In the context of design, much research has been carried out aimed at studying alternative materials. As well as being driven by innovation, DIY materials have as motivation for their development the designer's will to propose resources deriving either from the biological context or from the world of waste (DIY Materials, 2020). Designers are autonomously suggesting ideas for more sustainable and circular materials trying to motivate industrial production to invest in this area. Ideas are not lacking; what is lacking is more to do with the feasibility and economic convenience to encourage companies. It would seem that environmental problems alone and catastrophic predictions concerning life on Earth are not enough.

Research at a European level, on the other hand, has been active in this area, for example, in financing the recently concluded project, MaDe (2020). The project aimed at boosting talents toward circular economies across Europe, by partnering with design and cultural institutions. It focused on the figure of the material designer as an agent of change: designing, redesigning, reforming, reusing, and redefining materials to give them an entirely new purpose. Increasing the potential of materials, material designers can go on to research, advise, educate, and communicate what materials are and can be in the immediate, near, and far future, implementing positive social, economic, political, and environmental change across all sectors toward a responsibly designed future.

1.3 Where next?

Materials experience has helped designers and researchers transform materials and design from a narrow-focused technically oriented selection process into a vibrant, investigative, and inspirational practice. It provides a lens through which human–material relations may be examined at multiple levels of experience. While the expanding territories define the current state-of-the-art, an inevitable question remains: where next for materials experience? We offer some reflections on areas that we have started to make progress in, and which are likely to gain prominence.

- New forums and means of sharing DIY-Materials and their experiences will be established.
- MDD will be explicitly integrated with circular thinking, likely requiring renewed models, frameworks, and practices.
- MDD will be provided with extensions to support scaling-up of novel materials, mainstream adoption, and increased impact.
- Design education will embrace a hybrid approach of learning materials through community-generated information and by practical, experiential investigation and material creation.

What we are certain of is that materials experience will continue to feed the conscientious mind on the topic of materials and design: from a catalyst for new practices, through to an influencer of thoughtful choices of materials for design, as well as thoughtful decisions in new materials development, which will take into account our experiences with and through materials in tailoring material qualities. We foresee more intense blurring of the boundaries between material and product.

References

Ashby, M., & Johnson, K. (2002). *Materials and design: The art and science of materials selection in product design*. Oxford: Butterworth-Heinemann.

Ayala-Garcia, C., Rognoli, V., & Karana, E. (2017). Five kingdoms of DIY-Materials for design. In *Proceedings of EKSIG 17: Alive. Active. Adaptive—Experiential knowledge and emerging materials* (pp. 222–234). Rotterdam: Het Nieuwe Instituut/Delft University of Technology.

Barati, B., & Karana, E. (2019). Affordances as materials potential: What design can do for materials development. *International Journal of Design*, *13*(3), 105–123.

Camere, S., & Karana, E. (2018). Fabricating materials from living organisms: An emerging design practice. *Journal of Cleaner Production*, *186*, 570–584.

Collet, C. (2013). *This is alive*. Available at <http://thisisalive.com> Accessed 02.01.21.

DIY-Materials. (2020). Available at <http://www.diymaterials.it> Accessed 13.12.20.

Ferrara, M., Rognoli, V., Arquilla, V., & Parisi, S. (2018). Interactive, connected, smart materials: ICS materiality. In W. Karwowski, & T. Ahram (Eds.), *Intelligent human systems integration—IHSI 2018—Advances in intelligent systems and computing* (Vol. 722, pp. 763−769). Cham: Springer.

Franinović, K., & Franzke, L. (2015). Luminous matter electroluminescent paper as an active material. In *Proceedings of design and semantics of form and movement* (pp. 37−47). Milan: Politecnico di Milano.

Giaccardi, E., & Karana, E. (2015). Foundations of materials experience: An approach for HCI. In *Proceedings of CHI 2015* (pp. 2447−2456). Seoul: ACM Press.

Ginsberg, A., & Chieza, N. (2018). Editorial: Other biological futures. *Journal of Design and Science, 4*. Available from https://doi.org/10.21428/566868b5.

Ion, A., & Baudisch, P. (2020). Interactive metamaterials. *Interactions, 27*, 88−91. Available from https://doi.org/10.1145/3374498.

Ion, A., Kovacs, R., Schneider, O., Lopes, P., & Baudisch, P. (2018). Metamaterial textures. In *Proceedings of the 2018 CHI conference on human factors in computing systems* (pp. 1−12), No. 336. https://doi.org/10.1145/3173574.3173910.

Karana, E. (2009). *Meanings of materials* [PhD thesis, Delft University of Technology].

Karana, E. (2020). *Still Alive: Livingness as a material quality in design*. Breda: Avans University of Applied Sciences, ISBN: 978-90-76861-61-6.

Karana, E., Barati, B., & Giaccardi, E. (2020). Living artefacts: Conceptualizing livingness as a material quality in everyday artefacts. *International Journal of Design, 14*(3), 37−53.

Karana, E., Barati, B., Rognoli, V., & Zeeuw Van Der Laan, A. (2015b). Material driven design (MDD): A method to design for material experiences. *International Journal of Design, 9*(2), 35−54.

Karana, E., Blauwhoff, D., Hultink, H. J., & Camere, S. (2018). When the material grows: A case study on designing (with) mycelium-based materials. *International Journal of Design, 12*(2), 119−136.

Karana, E., Giaccardi, E., & Rognoli, V. (2017). Materially yours. In J. Chapman (Ed.), *The Routledge handbook of sustainable product design*. London: Routledge.

Karana, E., Hekkert, P., & Kandachar, P. (2008). Materials experience: Descriptive categories in material appraisals. In I. Horvath and Z. Rusak (Eds.), *Proceedings of the international conference on tools and methods in competitive engineering* (pp. 399−412). Delft: Delft University of Technology.

Karana, E., Pedgley, O., & Rognoli, V. (Eds.), (2014). *Materials experience: Fundamentals of materials and design*. Oxford: Butterworth-Heinemann.

Karana, E., Pedgley, O., & Rognoli, V. (2015a). On materials experience. *Design Issues, 31*(3), 16−27.

Karana, E., Pedgley, O., Rognoli, V., & Korsunsky, A. (2016). Emerging material experiences. *Materials and Design, 90*, 1248−1250.

MaDe (2020). Available at <http://materialdesigners.org> Accessed 13.12.20.

Manzini, E. (1986). *The material of invention: Materials and design*. Milan: Arcadia Edizioni.

Manzini, E. (1996). Design dei materiali. In A. Branzi (Ed.), *Il design italiano 1964−2000*. Milano: Mondadori Electa.

Manzini, E., & Petrillo, E. (1991). *Neolite-metamorfosi delle plastiche*. Milano: Edizioni Domus Academy.

Miodownik, M., & Tempelman, E. (2013). Light touch matters: The product is the interface. In. In R. Rodriques, & F. Pastonesi (Eds.), *Material matters VII* (pp. 30–35). Milan: Galli Thierry.

Myers, W. (2012). *Biodesign: Nature, science, creativity*. High Holborn: Thames & Hudson.

Parisi, S., & Rognoli, V. (2021). Design for ICS materials: The development of tools and methods for the inspiration and ideation phases. In V. Rognoli & V. Ferraro (Eds.), *ICS materials: Interactive, connected, and smart materials*. Milan: Franco Angeli/Design International.

Parisi, S., Rognoli, V., & Sonneveld, M. (2017). Material tinkering: An inspirational approach for experiential learning and envisioning in product design education. *The Design Journal, 20*(Suppl. 1), S1167–S1184. Available from https://doi.org/10.1080/14606925.2017.1353059.

Pedgley, O., Rognoli, V., & Karana, E. (2016). Materials experience as a foundation for materials and design education. *International Journal of Technology and Design Education, 26*, 613–630. Available from https://doi.org/10.1007/s10798-015-9327-y.

Rognoli, V. (2004). *I materiali per il design: Un atlante espressivo-sensoriale [Materials for design: An expressive-sensorial atlas]* [PhD thesis, School of Design, Politecnico di Milano].

Rognoli, V. (2020). Dynamism as an emerging materials experience for ICS materials. In V. Ferraro, & A. Pasold (Eds.), *Emerging materials and technologies: New approaches in design teaching methods on four exemplified areas* (pp. 92–104). Milan: Franco Angeli/Design International, ISBN: 9788835104513.

Rognoli, V., Bianchini, M., Maffei, S., & Karana, E. (2015). DIY materials. *Materials and Design, 86*, 692–702.

Rognoli, V., & Parisi, S. (2021). Material tinkering and creativity. In L. Cleries, V. Rognoli, S. Solaki, & P. Llorach (Eds.), *Material designers: Boosting talent towards circular economies*. <http://materialdesigners.org/>.

Trini Castelli, C., & Petrillo, A. (Eds.), (1985). *Il lingotto primario: Progetti di design primario alla Domus Academy*. Milano: Arcadia.

Veelaert, L., Du Bois, E., Moons, I., & Karana, E. (2020). Experiential characterization of materials in product design: A literature review. *Materials and Design, 190*, 108543. Available from https://doi.org/10.1016/j.matdes.2020.108543.

CHAPTER 2

How new materials speak: analyzing the language of emerging materials in architecture

Blaine Brownell
University of North Carolina at Charlotte, Charlotte, United States of America

2.1 Introduction

In the middle of crowded Hong Kong Island lies the former Central Police Station (CPS), Victoria Prison, and Central Magistracy complex. Once a prominent feature on the central Hong Kong hillside, the CPS is now dwarfed by high-rise towers on all sides. In recent years, the complex has been transformed into the Tai Kwun Centre for Heritage and Arts, a cultural destination and urban oasis (Tai Kwun Centre, 2018).

Upon first visiting the site, a visitor will notice a couple of unusual additions situated among the original 19th-century colonial buildings. Designed by Swiss architects Herzog & de Meuron (2018), the new structures—called the Old Bailey Wing and Arbuthnot Wing—stand in stark contrast with their historic surroundings (Fig. 2.1). Unlike the planar, masonry walls of stone and brick, the new wings are clad in a unique system of cast aluminum modules. According to the architects, "The unit system references existing masonry block elements on site in terms of scale and proportion," such as the existing granite retaining wall (Herzog & de Meuron, 2018).

Yet the layperson may not recognize this visual connection at all. Despite the proportional allusion, the new system appears completely alien: lozenge-shaped ellipsoids protrude at various depths and angles, their horizontal slits pointing outwards and at the streetscape below (Fig. 2.2). This thick aluminum screen partially obscures glazing that sits behind the primary surface, revealing hints of activity from within the building. Rather than turning the corner, the modules are cut at acute angles, revealing their intricate sectional profiles in a way that appears intentionally unfinished. Thus despite the architect's claims, the new system is wholly unlike brick or stone. It is hung rather than stacked, porous rather than opaque, and formally complex rather than simple. Furthermore, it is composed of a material and a set of details not seen in masonry walls.

Materials Experience 2. DOI: https://doi.org/10.1016/B978-0-12-819244-3.00009-0

FIGURE 2.1
Herzog and de Meuron, Old Bailey and Arbuthnot Wings, Tai Kwun Centre for Heritage and Arts. View showing historic context.

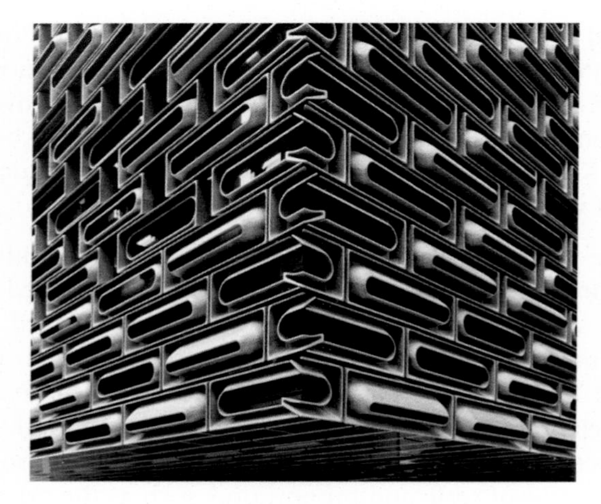

FIGURE 2.2
Herzog and de Meuron, Old Bailey and Arbuthnot Wings, Tai Kwun Centre for Heritage and Arts. Detail of building modules turning the corner.

One's initial confrontation with the curious facades of the Old Bailey and Arbuthnot Wings is representative of initial encounters with emergent materials and systems in general. When a visual distinction is made apparent, as is the case with many emergent materials, the human observer experiences several perceptual shifts. Responses may include confusion, intrigue, fascination, surprise, and delight. Architects like Herzog and de Meuron anticipate and plan for these reactions. In fact, the intent to shift public perception—and with it, visual culture—is a primary driver of their work.

But why is this case? What motivates architects and designers to transform users' attitudes via emerging materials and applications? How do they accomplish this objective, and what knowledge of material interpretation do they employ in the design process? Why are emerging materials of particular interest for architects and designers, as opposed to conventional materials? How might a broader understanding of emerging materials' perceptual potency, a fundamental type of material agency, enhance design praxis in general? This chapter will attempt to address these and other questions, focusing primarily on new materials and applications in building construction.

2.2 Emerging materials

First, what constitutes an emerging material? According to the Oxford English Dictionary, "emerging" or "emergent" is defined as "to come forth into view" and "to rise into notice, to come forth from obscurity." Another definition considers "the production of a type by such a process as evolution" (Oxford English Dictionary, 2020). Of note is a now obsolete connection between "emergent" and "emergency," "an unforeseen occurrence, a contingency not specially provided for" (Oxford English Dictionary, 2020). Yet another connotation is "an effect produced by a combination of several causes, but not capable of being regarded as the sum of their individual effects"—a scientific use that contrasts with "resultant," in which the outcome is predictable (Oxford English Dictionary, 2020). Thus for our purposes, the general meaning of "emerging/ent" connotes something in the process of becoming in a way that cannot be predicted.

At the time of this writing, there is no universally accepted definition of emerging materials. In the science and engineering fields, where the term is commonly used, the phrase refers to advanced materials. According to South Africa's Department of Trade and Industry (2019), "advanced materials refer to all new materials and modifications to existing materials to obtain superior performance in one or more characteristics that are critical for the application under consideration." The Versailles Project on Advanced Materials and Standards, established at the 1982 G7 Economic Summit, sought to address "the growing need for pre-standards research necessary to provide the technical basis for drafting specifications and codes of practice for advanced materials" (Early & Rook, 1996). The motivation concerned the "special properties" and "advanced processing procedures" that distinguish such materials and thus merit further scrutiny (Early & Rook, 1996). Today, advanced materials appear in journals such as Emergent Materials (2018), which claims to publish articles related to material research "at the forefront of physics, chemistry, biology, and engineering." Advanced materials are also the subject of programs such as the Argonne National Laboratory's Emerging Materials (2020), "a physics-based materials discovery program, emphasizing design, synthesis, and understanding of new materials..."

For the architecture and design fields, the term emerging materials is more inclusive, referring to both scientific and artistic and other forms of advancement. For example, the Emerging Materials program at T.U. Delft (2020) considers "novel, superior materials as better alternatives to convention." In design, emerging denotes one or more kinds of innovation, such as technological, functional, environmental, or esthetic. The implicit message is one of promise or potential: emerging materials represent not only individual achievements but also possible broader trends. The economist Clayton M. Christensen coined the phrase "disruptive technology" to describe this phenomenon from a business perspective. Simply put, a disruptive technology is a novel product or technique capable of displacing established technologies based on competitive functionality, price, or other characteristics (Christensen et al., 2004). Due to the nature of architecture and design as visually oriented fields, disruptive technologies may satisfy esthetic as well as functional and economic criteria.

For the purposes of this chapter, the term "emerging materials" will be used broadly—including references to advanced materials and disruptive technologies—but with an emphasis on esthetic and application novelty. Because the fundamental concern here is human experience, the examples to be discussed are commonly exposed to the human eye and other sensory organs, at a comprehensible scale. These materials' emerging qualities should be readily recognizable even if they are not themselves identifiable. Furthermore, in the spirit of the open-ended and multiscalar meaning of materials in design, the term "emerging materials" will be used inclusive of materials, products, systems, and assemblies.

2.3 The language of materials

Material language bears similarities with written and spoken language. According to novelist Victor Hugo (1964), "Architecture began like writing. It was first an alphabet. A stone was planted upright to be a letter and each letter became a hieroglyph. And on every hieroglyph there rested a group of ideas, like the capital of a column. Thus primitive races of the same period 'wrote' all over the world" (p. 175).

Throughout architectural history, materials have been used to represent significant ideas. Particularly before the invention of the printing press in the middle ages, architecture was expected to convey substantial moral and spiritual information to the public. With the birth of the Industrial

Revolution—after print media had become the predominant vehicle for information—architecture began to emphasize the intrinsic character, structure, and process of materials. In the early 20th century, the concept of authenticity gained importance in an architectural community obsessed with the establishment of a new, modern material language. "Let us be led by this enthusiasm which animates us," wrote Le Corbusier et al. (2007), p. 45 in "The Expression of the Materials and Methods of our Times ... Industrialization, standardization, mass production, all are magnificent implements; let us use these implements."

However, the material language of architecture has never entirely rid itself of the past. As architectural phenomenologist David Leatherbarrow and architect Mohsen Mostafavi argue in their seminal book *Surface Architecture* (2002), "Production and representation are in conflict in contemporary architectural practice." Buildings that prioritize the adoption of new technologies without thought to their expression abandon the project of representation, whereas projects that emphasize historical symbolism at the expense of contemporary building methods neglect the project of production (Leatherbarrow & Mostafavi, 2002, p. 1). When one considers the array of building products available today, one can see this phenomenon clearly evident at the material scale.

Based on this chapter's opening reference to masonry, it is pertinent to start with the common brick. The architect Louis Kahn advocated a "highest and best use" approach to building materials, and the brick is no exception. His support for construction appropriateness is evident in his imagined dialog with brick, given in a lecture in 1973: "You say to brick, 'What do you want, brick?' And brick says to you, 'I like an arch.' And you say to brick, 'Look, I want one too, but arches are expensive and I can use a concrete lintel over you, over an opening.' And then you say, 'What do you think of that, brick?' Brick says, 'I like an arch.'" (quoted in Twombly, 2003, p. 271).

The public has largely subscribed to Kahn's view of how brick should be used: to form load-bearing, compression-based structures composed of individual (and presumably handlaid) units. This widespread consensus is apparent in the way that brick is commonly represented. However, the reality is that brick—or "brickness" is typically a cosmetic treatment today, particularly in multistory and commercial structures. Thin brick-faced precast, glue-on brick, foam disguised as brick, wallpaper brick, and other such perversions maintain the pretense of brick without its material reality. To construct a direct link to written or verbal language: if one imagines a linear gradient of truth, one can begin to rate brick products on their authenticity concerning material ingredients, units, and uses (Fig. 2.3). Traditional brick would score high, brick wallpaper low, and glue-on brick somewhere in-between (based on its use of faithful ingredients).

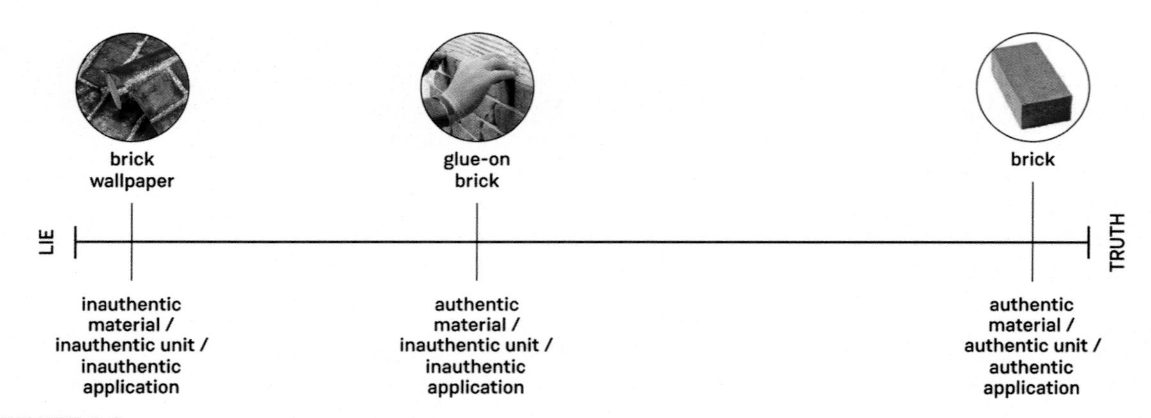

FIGURE 2.3
Gradient of truth: from brick to brick wallpaper.

2.4 Establishing a dialog

These aspects of materials—ingredients, units, and uses—describe physical details but not affects. As architects Farshid Moussavi and Michael Kubo contend, architects often experiment with the communicating potential of materials to influence perception. "These affects may start with found imagery or iconography as raw cultural material," they explain in The Function of Ornament (Moussavi & Kubo, 2008, pp. 7–8). "However, they do not remain as pure acts of consumption, but rather are disassembled and reassembled to produce new sensations that remain open to new forms of experience. It is in this way that they are contemporary and committed to progress." (Moussavi & Kubo, 2008, pp. 7–8) The creation of "new sensations" and "new forms of experience" require establishing an intentional dialog with the user. The many inauthentic materials available today—such as foam brick, fiberglass columns, or vinyl siding—are not intended to generate a dialog; they are meant to be ignored. If they are ever scrutinized, these materials are effectively camouflaged by fictive representations. In contrast, the affects described by Moussavi and Kubo are meant to be recognized by users, regardless of their authenticity.

A good example is Jun Aoki's Aomori Museum of Art in Aomori, Japan. Aoki conceived this project as a dialog between two primary material identities, one relating to excavated earth and the other to the snow-covered landscape aboveground. The courtyards and tunnels recessed below grade are composed of earthen walls and floors; the aboveground surfaces are clad primarily in white brick. From a distance, this brick resembles a standard running bond surface. Seen closer, however, it reveals a strange affect. The brick wraps the upper volumes comprehensively, including vast expanses of flat, cantilevered soffit (Fig. 2.4).

FIGURE 2.4
Jun Aoki, Aomori Museum of Art. View of brick soffit.

FIGURE 2.5
Jun Aoki, Aomori Museum of Art. Detail of brick turning corner.

One seldom sees brick applied in this way as it adds a large amount of unnecessary weight (the closest approximation of "appropriate" brick use would be in a shallow vault; otherwise, a lighter material would be used). Viewed more closely, the details of the building become even odder. With brick masonry walls, one expects to see the depth of the brick as it turns a horizontal or vertical corner. However, this is not the case in the Aomori Museum, where the corner details reveal the brick to be seeming without depth (Fig. 2.5). The architect has obviously specified an ultrathin brick for use as a veneer. This application resembles the everyday use described above, but with a notable exception: Aoki is intentionally showing us the thinness of the brick, not trying to disguise it. The interpretation is that he is expressing "the materials and methods of our times," as Le Corbusier described it, in an ironic way. He is revealing the fact that brick today is a decorative element.

Aoki's brick at Aomori clearly contrasts with Herzog and de Meuron's "new masonry." The materials, uses, and affects are obviously quite different. What unites them, however, is their shared aspiration toward affect—to shape the user's conscious experience of the material world. There are perhaps limitless ways to manifest such an affect in building materials, but a starting point is to propose three general strategies as a way to clarify preliminary design conceptualization. The first concerns the use of a common material or material language in an uncommon way. In its subversive details, Aomori fits this description. The second regards the application of an uncommon material to fulfill or represent a common application language. The Tai Kwun Centre corresponds with this approach. The third pertains to the use of an uncommon material in an uncommon application. Although relatively rare, such examples of holistic novelty do exist, and they convey their own particular affects.

2.5 Common materials, uncommon applications

When architects and designers consider the unconventional use of a standard material, they begin by pushing the conceptual boundaries of that material's expected function and performance. Karana et al. (2018) discuss the fine balance that exists between typicality and novelty in design and argue that the process of material acceptance involves more than considerations of utility. Studio Gang pursued this approach in the Marble Curtain project installed at the National Building Museum (Fig. 2.6). Designed for a 2003 exhibition entitled "Masonry Variations," the firm created an installation that questioned a seemingly irrefutable assumption: the fact that masonry may only be relied upon to handle loads under compression, not tension. Studio Gang worked with engineers to devise a tension-only application for marble. The design team developed modular, interlocking jigsaw pieces that were cut sufficiently thin by water-jet to transmit light. They then suspended these marble pieces from the ceiling with steel cables, adding more modules until the

FIGURE 2.6
Studio Gang Architects, Marble Curtain, National Building Museum, Washington, DC. *Studio Gang Architects.*

"curtain" reached the floor. Aside from a fiber-resin backing attached as a safety precaution, the stone operated entirely in tension without additional supports. From a distance, the installation conveyed the appearance of taut fabric; but up-close, observers were surprised to realize that the primary material was stone.

Another example installation is Vermilion Sands by Matthew Soules Architecture (Fig. 2.7). In this case, the shift involved thinking upside-down. Soules considered unexpected uses for geotextile fabric, which is intended to secure and improve soil structure, particularly along sloped or retaining surfaces. Rather than applying the fabric to the ground plane, however, Soules chose to elevate it. He created pyramid-shaped modules by wrapping the geotextile fabric around lightweight wire, then sprayed the modules with a slurry of ryegrass seeds and mulch—a process called hydroseeding. After allowing the grass to grow on the units in a nursery for 1 month, Soules inverted and suspended the modules from a steel frame, creating an expansive grass canopy overhead. The large structure incorporated water spray nozzles to mist the grass regularly and provide cooling for visitors on hot summer days. The result was an upside-down garden: a grass pergola that invited visitors seeking shade.

Another example, in this case a permanent construction is Swing Time by Höweler and Yoon Architecture (Fig. 2.8). The transformation here pertains to manipulating expected functionality. The primary material language in consideration is lighting [in this case, light-emitting diodes (LEDs)], which typically assumes the single function of illumination. In the spirit of multifunctionality, the architects combined lighting with the application of a playground swing set. With the charge to create a nocturnal playscape in a

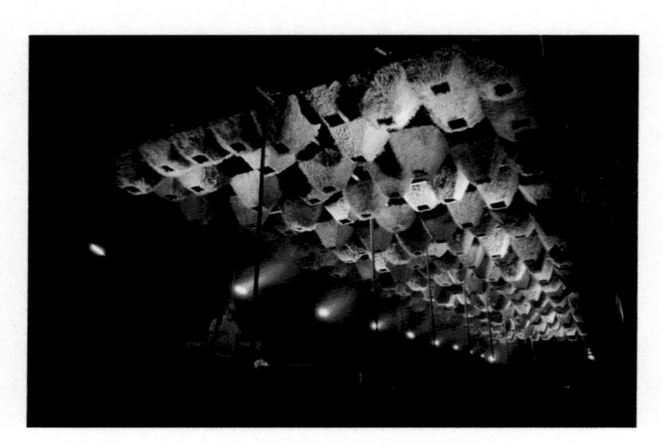

FIGURE 2.7
Matthew Soules Architecture, Vermilion Sands. *Sandy Wang.*

FIGURE 2.8
Höweler and Yoon Architecture, Swing Time. *Höweler and Yoon Architecture, LLP.*

dark and relatively unoccupied urban area, Höweler and Yoon set out to create an interactive, self-illuminating playground. They designed 20 custom swings made from welded polypropylene tubes, inside of which they placed LED lights, microcontrollers, and accelerometers. The color and intensity of the light vary based on how users interact with the swings. Slow activity prompts a soft white light, whereas fast swinging results in bright purple illumination. Not only does Swing Time demonstrate a novel use of lighting, but it minimizes the number of products required to create an illuminated play area.

2.6 Uncommon materials, common applications

Innovative material strategies can also emerge when the primary ingredients in everyday products, structures, or assemblies are transformed in significant ways. Returning to the brick theme, a notable example of this approach is BioBrick by North Carolina startup bioMASON (Fig. 2.9). As previously discussed, brick masonry is a nearly ubiquitous form of construction, which is used for a wide variety of residential and commercial applications. Traditional brick is made from formed, dried, and fired clay. The firing process lasts between 10 and 40 h, depending on the kiln, with temperatures up to 1300°C or higher (Brick Industry Association, 2006). Brick contributes about 0.2 kg CO_2 per kg of material produced—approximately double that

FIGURE 2.9
bioMASON, BioBrick. *bioMASON.*

of concrete (Ashby, 2013). bioMASON founder Ginger Dosier developed an alternative process, however, that employs bacteria instead of kilns. Channeling natural biomineralization processes, Dosier introduces *Sporosarcina pasteurii* to a mixture of yeast extract, urea, and calcium chloride in brick-shaped molds. After 5 days, the BioBrick—which resembles sandstone—is ready for use, with a strength equivalent to traditional brick.

Another common application that invites scrutiny is building insulation. Many of the various insulation products available today are derived from petroleum. Rigid polymer foams, such as polystyrene, contribute 3.7–4.1 kg CO_2 per kg of material. Some insulating foams, like extruded polystyrene, are made with hydrofluorocarbon blowing agents, which exhibit significant global warming potential (Ashby, 2013). Scientists at the Fraunhofer Institute for Chemical Technology identified an alternative material for building insulation: *Posidonia oceanica,* a seaweed known as Neptune grass. Mediterranean beaches are often lined with clumps of the so-called Neptune balls formed by washed-up seaweed, which local beautification crews often move to landfills. Recognizing Neptune grass' advantageous qualities, including moisture tolerance and rot resistance, the Fraunhofer researchers devised a way to dry the clumps and cut the material into collections of short fibers. These fibers require minimal processing and no chemical additives and may be treated as effective loose-fill insulation, to be blown into roof or wall cavities.

Another omnipresent application in need of rethinking is the highway barrier: specifically, the sound barrier wall erected to stop the noise and light pollution generated by vehicular traffic from spilling into adjacent residential neighborhoods. According to architecture firm Fieldoffice, there are over linear 75,000 km of highways lined with sound barrier walls in the United States alone (Hutchins, 2008). The firm's Superabsorber is designed as a replacement barrier system, composed of porous, titanium dioxide (TiO_2)-rich concrete. Emulating the form of a sea sponge to maximize surface area, the photocatalytic concrete reduces local air pollutants when activated by sunlight. Based on cement manufacturer Italcementi's demonstration that 15% of TiO_2-based cement coverage results in a 50% reduction in airborne pollution, the new barrier wall functions to reduce air pollution—as well as noise and light pollution—while representing a significant esthetic upgrade from typical sound barrier walls (Schelmetic, 2012).

2.7 Uncommon materials, uncommon applications

This strategy is the most novel of approaches, given the unconventionality of both the material and its application. The term "uncommon material" here refers to the product, unit, or module more than actual chemical ingredients—although these too can be novel in some cases. For example, the Pneumatic Biomaterials Deposition project by the Mediated Matter Group at the Massachusetts Institute of Technology (MIT) employs both a material and a procedure rarely used in building construction. Most building materials are rigid elements, and those that are soft, pliable, or fluid are typically contained by, within, or between rigid components. However, the MIT research demonstrates a significant departure from this norm. Looking at the living tissues of biological organisms, the researchers devised a method to print hydrogels via pneumatic deposition. They connected a custom-designed apparatus of various syringes and nozzles to a robotic arm, which enables liquid-based additive manufacturing with precise control. The machine can print fluids ranging in viscosity between 500 and 50,000 cPs, which transform into elastic solids after printing. Initial applications include experimental wearable prosthetics, tissue scaffolds, and soft architectural skins.

In the case of the Air Flow(er) project by Lift Architects, the uncommon material is shape memory alloy (SMA), which is utilized to impressive collective effect (Fig. 2.10). Air Flow(er) is a smart, responsive ventilating surface composed of individual modules resembling flowers. The system was inspired by the natural principle of thermonasty, as seen in the crocus flower, which enables plants to move when ambient temperatures change. Each module consists of four thin gusset plates attached to a frame by nickel- and titanium-based SMA cables. The SMA material automatically responds to trigger temperatures around 16°C and 27°C: when the air heats up to the higher

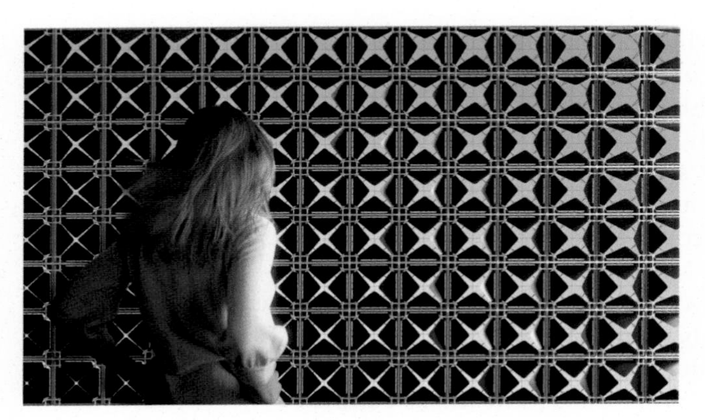

FIGURE 2.10
LIFT Architects, Air Flow(er). *Andrew O. Payne, LIFT Architects.*

FIGURE 2.11
Heatherwick Studio, Seed Cathedral, Shanghai, China. View at night showing illuminated rods.

temperature, the panels open; when it falls below the lower set point, the panels close. In this way, the system dynamically responds to the local climate by increasing porosity and ventilation when needed, all without electricity.

Another example of holistic innovation is the Seed Cathedral created by Heatherwick Studio (Fig. 2.11). Designed as the UK Pavilion for the 2010

World Expo in Shanghai, the temporary structure was sufficiently novel when built that it was difficult to describe. Composed of over 60,000 acrylic rods that projected at all angles, the pavilion appeared to be more of a cloud than a solid structure. When seen in person, the wind-responsive rods bobbed almost imperceptibly, conveying the effect of a continuously blurred image—a soft, thickened surface always in motion. Each 7.5-m-long acrylic element extended into the interior of the pavilion, resulting in a similarly delineated interior cavity. Heatherwick Studio embedded a different type of seed from the Kew Gardens Millennium Seed Bank Project at the end of each optical strand. Driving ambient light deep inside the Seed Cathedral, the tubes outlined the seeds in sharp silhouettes, thereby conceptually transforming an atmospheric construction into an earthbound seed bank.

2.8 Advancing material linguistics

The three material strategies outlined in Sections 2.5 through 2.7 are not static, but vary with context. The first two approaches depend on the conceptual starting point of the designer and may be interchangeable when the emphasis shifts between a material and its application. For example, the Marble Curtain project was initiated with a material question about the properties of stone. However, given the result, the designer could conceivably have started with the notion of a curtain as an application and asked what uncommon material could be used to realize such an application. In this case, the example would shift to the category of Uncommon Material, Common Application. Furthermore, the third method (Uncommon Material, Uncommon Application) is dependent upon an audience's perspective on its degree of novelty. For example, when LEDs were first introduced as a disruptive lighting technology, a project like Swing Time would have been classified in the third category. However, in time, LEDs have become a familiar technology with widespread commercial availability—thus landing the project in the first classification. In other words, the third strategy of Uncommon Material, Uncommon Application is highly subjective, based on the anticipated users' familiarity with both aspects of material language.

In all of these cases, the fundamental question concerns the frame of reference. The attributes that determine whether a material or an application is common or uncommon relate to the general concept of baseline. According to Merriam-Webster, baseline is "a usually initial set of critical observations or data used for comparison or a control" and "a starting point" (Merriam-Webster, 2020). For our purposes, the notion applies to commonality in availability and practice. Baseline knowledge is accumulated through observation as well as industry experience. In simplistic terms, clay bricks and wood doors would be considered baseline, or standard, building materials, whereas the inverse—wood bricks and clay doors—would not. That said, baseline varies with culture, geographic location, time, and other factors. For example, asphalt shingles are considered a baseline roofing material in the United States, but not in Europe; residential carpeting is likewise commonplace in the United States, but not in China. Notably, baseline is also a moving target: what is considered typical at a given moment and location inherently changes as new technologies and standards are introduced and adopted.

For architects and designers, it is critical to have a keen sense of baseline to impart meaningful material affects to observers. The aim should be to

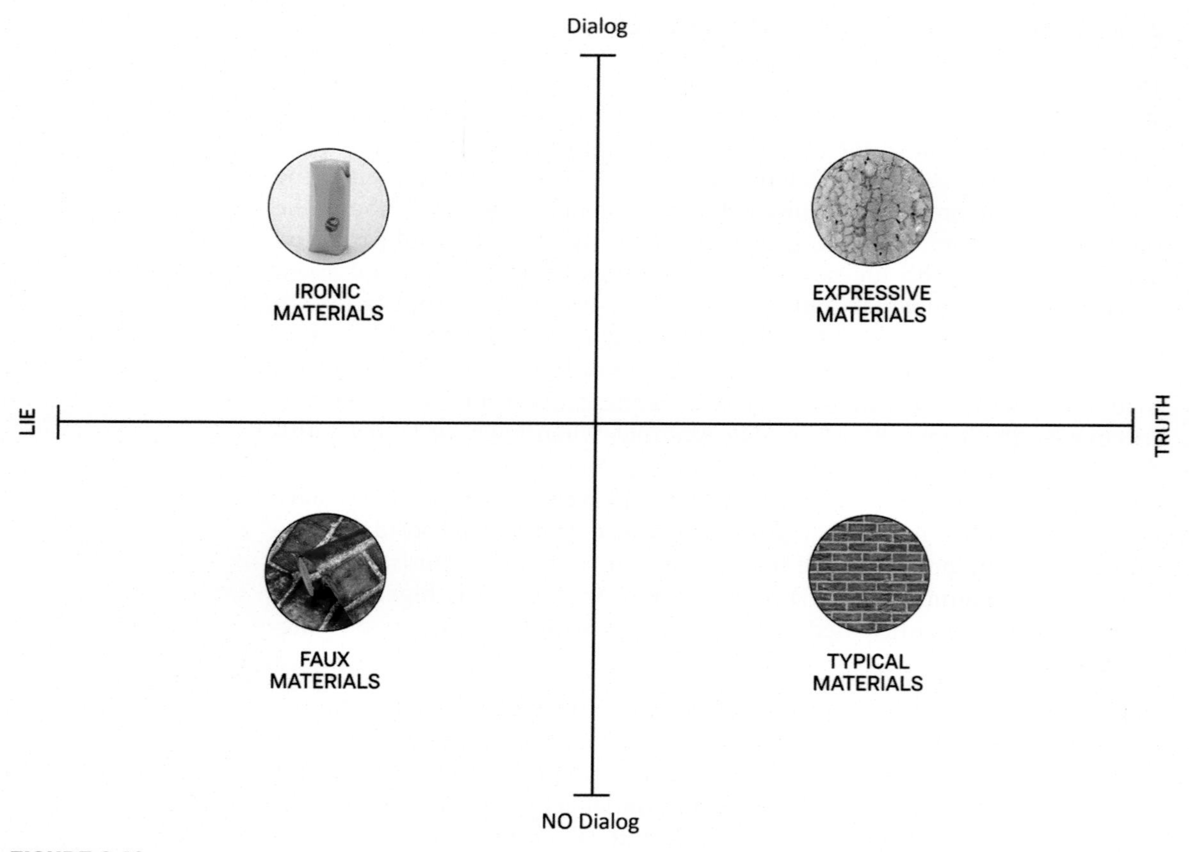

FIGURE 2.12
Quadrant diagram.

connect with the user consciously, with an uncommon material, application, or both, to establish a dialog of material language. Such an approach demonstrates respect for users, acknowledging that they have an awareness of conventional materials and practices shaped by experience and that they can appreciate conceptually rich departures from the baseline. The nature of this acknowledgment can vary as directed by the designer. For instance, the new brick language in Jun Aoki's Aomori Museum of Art is subtle and likely unnoticed by most visitors, whereas the otherworldly visage of Heatherwick Studios' UK Pavilion is visually arresting for most audiences. Nevertheless, the ambition to create—and achieve—a dialog with the user is a fundamental differentiator between innovative and conventional material language. Fig. 2.12 transforms the linear gradient of "truth" in Fig. 2.3 into a quadrant diagram with a second gradient called "dialog" (Fig. 2.12). In this way, one could argue that the traditional view of materials is that they stay silent; meanwhile, a contemporary perspective is that they are communicating vehicles (Fig. 2.13 and 2.14). Such a paradigm elevates material language to a higher place within the broad spectrum of human communication and accepts that the designed environment is at least as significant a linguistic milieu as literary and oratory spheres.

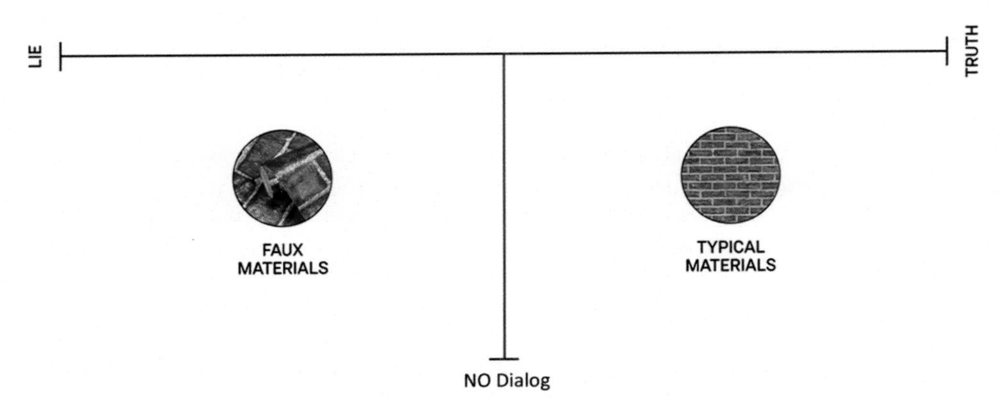

FIGURE 2.13
Quadrant diagram: silent matter.

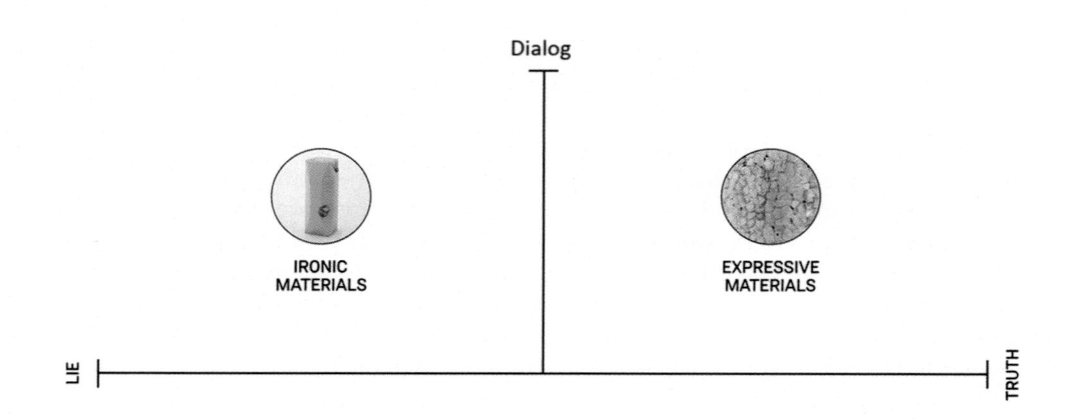

FIGURE 2.14
Quadrant diagram: communicative matter.

References

Argonne National Laboratory (2020). *Emerging materials group*. <https://www.anl.gov/msd/emerging-materials> Accessed 30.05.20.

Ashby, M. (2013). *Materials and the environment: Eco-informed material choice* (2nd ed.). Amsterdam: Elsevier.

Brick Industry Association (2006). *Manufacturing of brick*. <http://www.gobrick.com/docs/default-source/read-research-documents/technicalnotes/9-manufacturing-of-brick.pdf?sfvrsn = 0> Accessed 30.05.20.

Christensen, C., Anthony, S., & Roth, E. (2004). *Seeing what's next: Using the theories of innovation to predict industry change*. Cambridge: Harvard Business Review Press.

Department of Trade and Industry, Republic of South Africa (2019). <http://www.thedtic.gov.za/sectors-and-services-2/industrial-development/advanced-materials/> Accessed 30.05.20.

Early, J., & Rook, R. (1996). Versailles project on advanced materials and standards (VAMAS). *Advanced Materials*, *8*(1), 9–10.

Emergent Materials (2018). <https://www.springer.com/materials/journal/42247> Accessed 30.05.20.

Herzog, J., & de Meuron, P. (2018). <https://www.herzogdemeuron.com/index/projects/complete-works/276-300/296-tai-kwun.html> Accessed 30.05.20.

Hugo, V. (1964). *The hunchback of Notre-Dame* (W. Cobb, Trans.). New York: Signet (Original work published 1831).

Hutchins, S. (2008). Excerpts from the techno-bible, *Architect*, May 22. <https://www.architect-magazine.com/technology/excerpts-from-the-techno-bible_o> Accessed 30.05.20.

Karana, E., Blauwhoff, D., Hultink, E.-J., & Camere, S. (2018). When the material grows: A case study on designing (with) mycelium-based materials. *International Journal of Design, 12*(2). Available from http://www.ijdesign.org/index.php/IJDesign/article/view/2918/823, Accessed 30.05.20.

Le Corbusier. (2007). Architecture: The expression of the materials and methods of our times. In W. Braham, J. Hale, & J. S. Sadar (Eds.), *Rethinking technology: A reader in architectural theory*. London: Routledge. (Original work published 1929).

Leatherbarrow, D., & Mostafavi, M. (2002). *Surface architecture*. Cambridge: MIT Press.

Merriam-Webster (2020). *Definition of 'baseline.'* <https://www.merriam-webster.com/dictionary/baseline> Accessed 30.05.20.

Moussavi, F., & Kubo, M. (2008). *The function of ornament*. Barcelona: Actar.

Oxford English Dictionary (2020). *Definition of 'emergent.'* <https://www.oed.com/view/Entry/61126?rskey = OkB1oC&result = 1#eid> Accessed 30.05.20.

Schelmetic, T. (2012). Titanium dioxide coats buildings, structures to help them stand up to Smog Monster. Thomasnet.com, May 22. <https://www.thomasnet.com/insights/imt/2012/05/22/titanium-dioxide-coats-buildings-structures-to-help-them-stand-up-to-smog-monster/> Accessed 30.05.20.

T.U. Delft (2020). *Emerging materials program*. <https://www.tudelft.nl/en/ide/about-ide/departments/sustainable-design-engineering/research-areas/emerging-materials/> Accessed 30.05.20.

Tai Kwun Centre (2018). <https://www.taikwun.hk/en/> Accessed 30.05.20.

Twombly, R. C. (2003). *Louis Kahn: Essential texts*. New York: W.W. Norton & Company.

CHAPTER 3

Experiential craft: knowing through analog and digital materials experience

Nithikul Nimkulrat
OCAD University, Toronto, Canada

3.1 Introduction: crafting with analog and digital materials

This chapter investigates the intertwinement of analog and digital materials experience generated from within a process of handcrafting through digital means. The work presented in the chapter is fundamentally an exploration of the relationship between the maker (i.e., the author), materials, and tools, or "craft," carried out within the research *through* design framework (Frayling, 1993). In this exploration, the author's design intentions reside in the hands executing her craft with judgment, care, and skill (Pye, 1968). Craft here is a means of thinking logically through sensing and immediate experience of materials (Nimkulrat, 2010, 2012) and "a dynamic process of learning and understanding through material experience" (Gray & Burnett, 2009, p. 51). Such an experiential process entails "intimacy and affinity to human values and emotion and the ability to experiment and subvert" which are the essential characteristics and strength of craft (Niedderer & Townsend, 2014, p. 631). In addition, by incorporating digital fabrication and tools in craft processes, the work extends Adamson's (2007) notion of craft as "a way of thinking through practices of all kinds" (p. 7).

Craft is generally concerned with "controlling the whole process from start to finish, adopting, adapting and improving tools as the need arises," while the process of working with digital technologies seems "'hidden' making understanding and controlling the process from concept to end product seem more complicated … and not craft" (Shillito, 2013, p. 9). This raises questions about the role of the controlling hand in *machine culture* and Computer-Aided Design (CAD) environments of industrial design where mechanized output is associated with ideas of precision, certainty, and reproducibility. "[A]gainst the rigorous perfection of the machine, the craftsman became an emblem of human individuality, this emblem composed concretely by the positive value placed on variations, flaws, and irregularities in handwork" (Sennett, 2008, p. 84). Pye (1968) relates craft to risk-taking:

Materials Experience 2. DOI: https://doi.org/10.1016/B978-0-12-819244-3.00026-0

"[Craftsmanship] means simply workmanship using any kind of technique or apparatus, in which the quality of the result is not predetermined,... The essential idea is that the quality of the result is continually at risk during the process of making" (p. 2). Traditional craftspeople make personal subjective decisions while working with analog materials to form artifacts and generally have no digital history to retrace their decisions, that is, the only documentation is material artifacts (Zoran & Buechley, 2013, p. 6). Today's craftspeople gain access to digital fabrication and tools, designing an object on a computer and then fabricating it by a machine. Craft as expanded by digital media has the capacity to "reunite visual thinking with manual dexterity and practiced knowledge" (McCullough, 1996, p. 50). With their right approach, skills, and mindset, craftspeople can find the connection between handcrafting and digital technologies (Campbell, 2007). The subjective decisions of the maker are still indispensable for the production of digitally fabricated artifacts, which in turn reflect their maker's skills, perspectives, and values (McCullough, 1996). Like handcrafted work, digitally fabricated work is at risk and its result as a material artifact acts as the documentation of the process. Yet, as the digital craft practitioner has access to editable digital files that are a rich history of the process, it can be implied that digital craft is at significantly less risk than traditional analog craft.

3.2 Three concepts of materials experience: materialness, material-driven design, and material agency

In order to examine analog and digital materials experience through own craft practice, it probably is necessary first to grasp a broad subject matter of materials experience. This section therefore aims to ground an understanding of the relationship between materials (i.e., the substance of artifacts) and experience (i.e., a way to gain knowledge about the world), or how materials make us think, feel, and act (Giaccardi & Karana, 2015). Several concepts regarding materials and their role in forming meaningful objects have recently emerged, particularly in the disciplines of design, archeology, anthropology, and cognitive science, for example, *materialness* (Nimkulrat, 2009), *material-driven design (MDD)* (Karana, Barati, et al., 2015), and *material agency* (Gell, 1998; Knappett & Malafouris, 2008). While these concepts differ from one another in their detail and disciplinary contexts, the key point shared among them is the material's active role in the artifact and its relation to people.

Over a decade ago, the author's practice-led research into the expressivity of physical materials from a textile practitioner's experiential perspective proposed the concept of *materialness* for the design process of material-based craft (Nimkulrat, 2009). Developed from Dewey's (1934) notion of *art as experience* and Heidegger's (1962) and Merleau-Ponty's (1962) phenomenology, the concept reveals the entirety of the creation that is rooted in a material, including the elements of form, content, context, and time. Serving as an alternative means to assist textile practitioners in the creation of artifacts using any material, the concept formulates both the physical form and the subject matter of each artifact and its contextual application. In other words, the artifact is a formed material incorporating materiality and subjectivity. When a textile practitioner experiences a tangible material, she not only feels its physical properties, such as strength or weakness and roughness or smoothness, but also associates these qualities to her own expressive capacity. By placing an emphasis on a selected material according to the materialness concept, a textile practitioner can create the form and content of an artifact and bring the elements of context and time to her creation in order to design an overall experience for the viewer or user of the artifact. The physical qualities of a material affect the ways in which the maker and others understand objects. Through material engagement, a textile practitioner learns through senses how to manipulate the material at hand and is eventually able to improvise the manipulation technique to create an inimitable artifact that represents the maker's authenticity. The artifact thus becomes the physical embodiment of its maker's expressive thought; the creative and transformative act of creation embeds meaning in the material artifact and the artifact, in turn, defines and expresses its meaning through its physicality.

The materialness concept resonates with MDD, a method to design with and for material experiences, offered by Karana et al. (2015). The method is grounded on the accounts of materials experience (Karana et al., 2008; Karana et al., 2014, 2015) and experience design (Desmet et al., 2011). Similar to the materialness concept, the method suggests the departure of a product design process from materials and shifts the designer's role from a passive recipient to an active explorer. The MDD method involves four steps: (1) understanding the material; (2) creating materials experience vision; (3) manifesting materials experience patterns; and (4) creating material/product concepts (Karana et al., 2015, pp. 41–46). This is in contrast to the systematic approach of product designing, in which the formulation of problems and conceptualization of ideas would come first and then proceed to the translation of the concept into forms, functions, and materials embodied in a final design product (Cross, 2008). By utilizing the MDD method, both material and product are developed within a single design process, resulting in new material proposals and product application concepts.

Although the MDD method arises from industrial design research that differs from the disciplinary context of the materialness concept which is from textile craft research, both approaches are similar in their principle and focus on materials and design for experience. Both materialness and MDD aim to change the way designers work with materials. With the materialness concept, instead of beginning the design process by conceiving the idea of an artifact and sampling different types of material in order to find the right material whose qualities elicit the response that corresponds with the idea, the textile practitioner starts with material selection and lets the chosen material leads the creative process (Nimkulrat, 2009, pp. 229–230). The experiential qualities of the material as understood by the textile practitioner not only inform ways in which the concept and form of the artifact may be designed but also evoke possible manipulation techniques and tools to achieve the design intention. Likewise, the MDD method begins with understanding the material and results in material and product concepts blended in one final outcome of design for experience (Karana et al., 2015, p. 49).

In archeology, anthropology, and sociology, a notion that emphasizes active materials is *material agency*. While this concept is widely adopted in recent research in various fields of design, the concept of agency as such has been subject to several interpretations (Robb, 2010). One of the early accounts of agency is tackled from the perspective of anthropology of art; Gell (1998) considers art and artifacts as having "distributed agency" because they produce effects and cause people to feel—"Agency is attributable to those persons (or things...) who/which are seen as initiating causal sequences of a particular type, that is, events caused by acts of mind or will or intention, rather than the mere concatenation of physical events" (p. 16). Material objects according to Gell (1998) are positioned in social relations that are normally associated with persons, not with inanimate things, and can therefore be agents as long as humans interact meaningfully with them. On the contrary, Knappett (2005) argues that because objects are not alive they cannot have agency on their own, but they may act in a manner similar to that of an agent when they are imbued by humans with a purpose. The most relevant accounts of material agency to the design discipline probably are ones illustrated by Malafouris (2008) and Pickering (1995, 2010, 2013). Both see that agency emerges only when people and things are mutually related and interact with one another. Pickering (1995, 2010, 2013) proposes a way to understand natural sciences and knowledge production through a "dance of agency," a concept that deals with "actions, human or non-human, in the material world and the interplay of those" (Pickering, 2010, p. 195). In this dance, there is a practice, a performance, or a set of ongoing interactions between humans and nonhumans, for example, a scientist is "trying this, seeing what happens, trying something else" (Pickering, 2013, p. 81).

On the other hand, Malafouris (2008) brings together materials and people in the process of skilled practice in which action comes before intention and materials engage human agents in a dialog. He examines agency and intentionality as "the properties of material engagement, ... where brain, body and culture conflate" (Malafouris, 2008, p. 22) and suggests that they should be understood as a "distributed, emergent and interactive phenomenon rather than as a subjective mental state" (p. 33). His explanation through the case of pottery making in which the potter interacts with clay on a wheel highlights that the artifact should not be interpreted as a passive object of the maker's intentionality but as "a functionally cosubstantial component of the intentional character of the potting experience" (Malafouris, 2008, p. 33). Malafouris's (2008) point for discussing agency in relation to material engagement is to emphasize that agency is not a fixed attribute of either people or objects but the spatiotemporally relational and emergent product of material engagement existing as the flow of the situated, dialectic activity itself (p. 35). In his later work, Malafouris (2013) argues for the differentiation between agency and sense of agency in the context of "real-world experience" and "situated action" (p. 215). Taking the same example of pottery making, Malafouris (2013) points out that what differentiates the potter from the clay and the wheel is the potter's possession of *sense or experience* of agency, which enables her to be an "experiencing enactive agent ... conscious of being in control of one's own actions" (p. 215). The material (i.e., clay), the maker (i.e., potter), and the tool (i.e., wheel) are related in a material engagement situation (i.e., pottery making). This line of thinking will be taken on to illuminate the author's real-world experience of craft practice through analog and digital materials in the next sections.

3.3 Handcrafting through digital tools

This section describes the initial process of handcrafting through digital means involved in a research project that attempts to understand how digital tools can be used to translate handcrafted objects in craft practice. The project addresses two research questions: (1) What forms of knowing and meaning making evolve in analog and digital craft practice? (2) What does it mean to explore material in CAD through Virtual Reality (VR)? The exploration which lasted for 2.5 months in the Digital Fabrication and Mixed Reality labs at Emily Carr University of Art + Design in Canada is by nature concerned with both analog and digital materials experience.

Prior to this investigation, the author's craft practice had utilized neither tool nor machine for over a decade. Manipulating paper string by hand without any tool or machine would leave no boundary between the material and the maker. This form of practice was initiated with an intention of examining the expressivity or active role of material in creative processes (Nimkulrat, 2009).

FIGURE 3.1
(A) Hand-knotting process of a small artifact from paper string in 2017 and (B) the original cup form made in 2007.

In order to begin her new experience with digital processes through a craft lens, the author hand-knotted a small artifact from paper string for further experimentation with digital tools (Fig. 3.1).

3D scanning and 3D printing seemed to be the most straightforward process for translating the analog artifact into a digital format. Scanning using a high-definition 3D laser scanner required the moving hand and eye coordination to capture the rows and columns of knots (Fig. 3.2). Scans of the cup showed a line quality resembling the characteristic of paper string and the intricate handcraft. Nonetheless, the files of the scans were too large to further process effectively in CAD, causing computer crashes. This revealed that the properties and characteristics of the handcrafted object were beyond the capacity of this digital tool.

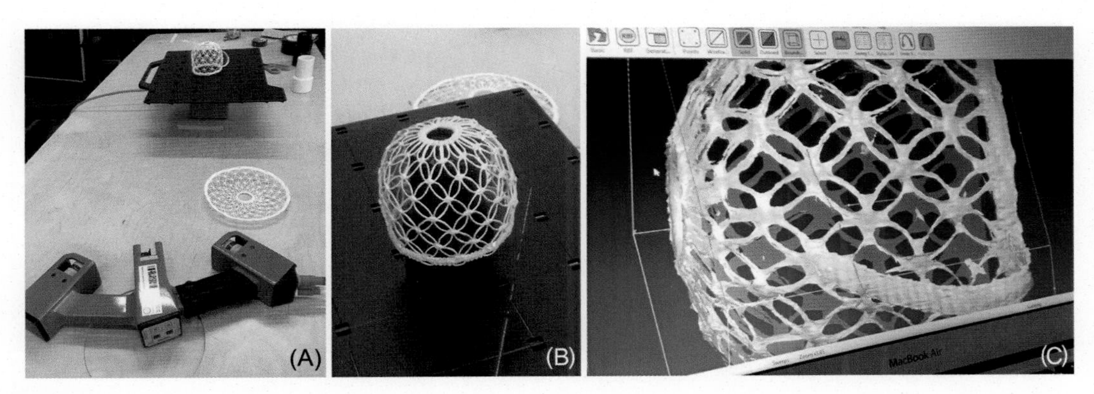

FIGURE 3.2
(A) Polhemus Scorpion handheld 3D laser scanner, (B) the scanning process, and (C) the 3D scan file.

Freehand drawing on a photograph of the analog artifact with a stylus on a tablet was the next approach adopted to produce a simplified model with manageable amount of data. This approach followed Campbell's (2016, p. xxi) idea of abstraction to simplify the degrees of complexity in working with digital tools. Handling a digital tool to interact with the CAD program resonates with Clark & Chalmers's (1998) *Extended Mind* thesis, which considers the mind not necessarily residing in the brain or physical body but having a capacity to extend itself to elements of the environment. In this case, the author's mind extended to the virtual software and her hand to the digital tool to construct a CAD model of a three-dimensional array of one section of knots that resembled the original hand-knotted cup (Fig. 3.3).

The work, to this point, was a record of materials experience according to analog parameters (i.e., things that string does well such as self-friction, knot, and bend) translated into a prescriptive CAD language according to the digital parameters of the software. Several hours were invested in the development of the 3D model and the navigation of the software's restrictions to achieve a model suitable for output. During this process, most of the author's experience was gained through the visual and the handling of digital tools but not so much through the tactile qualities of physical material. By engaging with the digital tools in the flow of this situated exploration, agency emerged here in the relational negotiation between the author's hand handling a tool and the visual on the tablet or the laser beam visible on the hand-knotted cup.

3.4 Material engagement in digital fabrication

With the author's little previous experience with digital fabrication, it was crucial for her to first gain a better understanding of the capacity of 3D-printing

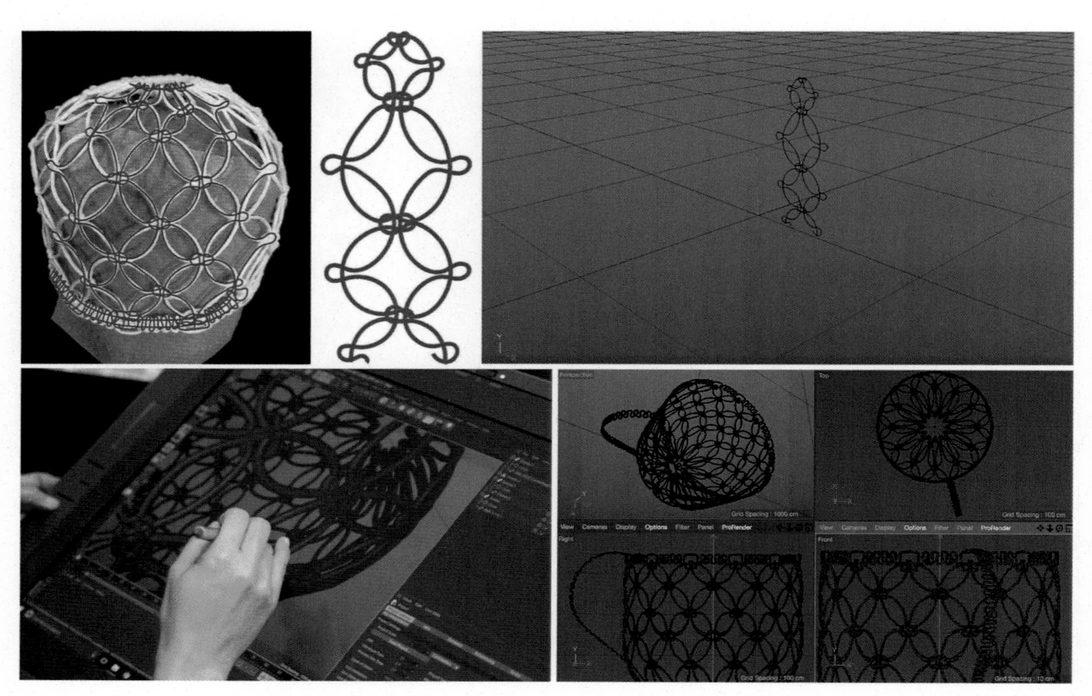

FIGURE 3.3
The process of 3D modeling of the knotted cup.

technology in relation to the delicate cup form. The most common process of material extrusion was adopted, using a Fused Deposition Modeling printer and polylactic acid (PLA) thermoplastic made from renewable resources (e.g., corn starch) and thus biodegradable.

To understand the limits and possibilities of the selected process, details of the CAD model were set as small as 0.4 mm and used only partial support material. The printer could print the whole model, but the physical print was too fragile to retain the cup form (Fig. 3.4).

In search for the thickness suitable for the capacity of the machine while preserving the characteristics of knots, likeness of strings, and fidelity of hand-knotting, the CAD model's thickness was increased to 0.8, 0.95, and 1.2 mm and test printed, respectively. "[B]y making something happen more than once, we have an object to ponder; variations in that conjuring act permit exploration of sameness and difference; practicing becomes a narrative rather than mere digital repetition" (Sennett, 2008, p. 160). The comparison of the resulting prints showed that the 0.95-mm print was the most successful rendering and would be used for further 3D printing using wood (approx. 30% wood fibers, 70% PLA) and

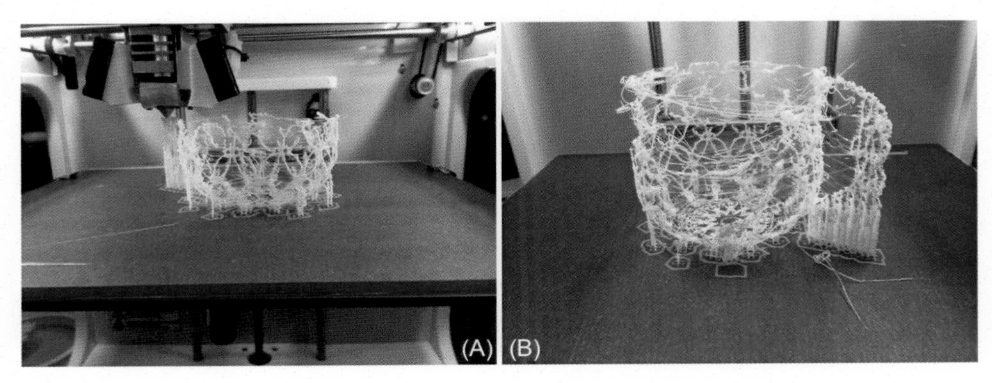

FIGURE 3.4
(A) The first 3D printing to test the limits and possibilities of the printer and (B) the resulting print.

copper (approx. 30% copper, 70% PLA) composite filaments, in addition to the standard PLA. Printing the 3D model of the cup using various composite filaments was expected to allow for a detailed comparison of the printed outcomes and a consideration of how intrinsic material characteristics might influence the form and meaning originated in the mediated artifact.

In the printing process, characteristics of natural components of wood and copper composites such as smell and texture were sensorially recognizable. To explore each composite's properties, several iterations of parameter settings were made in the printer's software. The printer's settings such as temperature, speed, density, and angle of support material were modified in order to find a solution to print each material successfully. As machines "demand a uniformity and standardization of material, not only in size but in density and consistency" in order to work efficiently (Risatti, 2007, p. 197), these composite materials are in a standard form and size compatible with the 3D printers. However, as the composites are partly made from natural components such as wood and copper, irregularities occurred and caused problems to the performance of the printer. To overcome the problems, further modification of the CAD model itself was required when no successful results were attained after the adjustment of parameters in the printer's software. For example, to solve the repeated clogging of the extruder nozzle caused by the wood composite's fibrous and burnable properties, the speed for printing was gradually increased and the temperature was decreased (from 220°C to 210°C) to achieve a better flow of filament. Despite the revised parameter settings, the resulting prints were still missing parts, hence calling for simplifying the model's geometry (Fig. 3.5). Although the author adopted neither the materialness concept nor the MDD method as such, she conducted these iterations of modeling and parameter setting by placing an emphasis on each material as to how its

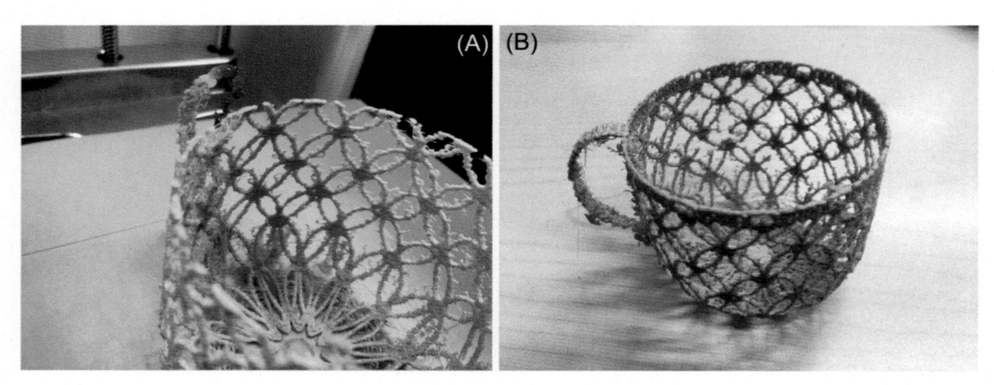

FIGURE 3.5
Printing of wood composite. (A) An incomplete print due to the clogging of the extruder nozzle and (B) a print with missing parts.

properties influenced the printing process. Again, this was a situated activity in which the maker negotiated with the material and the tools—a relational dialectic situation through which the agency of the material and the agency of the maker were enacted. The notion of agency here focused on the emerging properties of the materials and how they actively contributed to the way the design activity progressed; it was not a fixed property instilled in the artifact but enacted in the relationship between people and artifacts. Materials became actors instead of insignificant artifacts to be acted upon.

After several iterations of parameter settings, CAD geometry, and printing, the outcomes achieved a satisfying material conformity. Having observed the printed cups, the author noticed that each composite performed distinct material traits. The wood cup had a fibrous effect resembling the growth of roots and was much lighter than the ones printed using copper and PLA. The printed cups had their own characteristics as expressed by the material used and their appearance comparable to the original cup (Fig. 3.6). The composite filaments that looked similar to one another prior to extrusion could result in the artifacts with detailed features unique to the materials. Their true characteristics became recognizable once they were passed through a heated extruder nozzle, layer by layer according to the programmed models of the same cup form.

This material exploration has illustrated that the production of digitally produced artifacts still requires the maker's decisions (McCullough, 1996) as well as her judgment, care, and skill (Pye, 1968) similar to the production of handcrafted artifacts. This was due to the digital technologies being not as precise and certain as assumed. What was key here is craft as "knowledge that empowers a maker to take charge of technology" (Dormer, 1997, p. 140). Technology did not merely fulfill the author's creative intention of translating a handcrafted artifact into a digital format. Instead, her engagement with digital fabrication

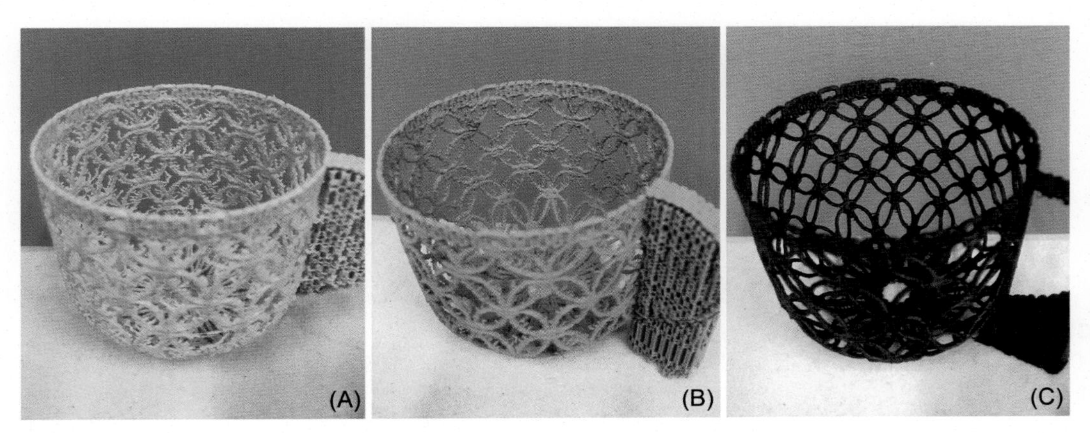

FIGURE 3.6
(A) Wood print, (B) copper print, and (C) PLA print.

tools and materials in the creative action enabled technology and human to inform each other. The material (i.e., a type of filament), the tools (i.e., CAD program; 3D printer and its software), and the maker were related in a material engagement situation (i.e., 3D modeling and printing).

3.5 Gestural crafting in virtual reality

Although the 3D-printed cups showed well the likeness of the original hand-knotted cup, the author felt that they missed the physical characteristics of knots, for example, continuity, flexibility, and bendability, or things that string does well. To represent the nature of knots, the next experimentation aimed to create a CAD model of flexible, loose knots. The same tools and methods—drawing with a stylus on a tablet—were employed to create a section of knot pattern for further 3D modeling. Notwithstanding the author's thirteen-year experience of hand-knotting three-dimensional artifacts, knotting virtually on a 2D screen was incomprehensible for her. She found that working on a 2D screen to create a 3D model did not amply illustrate or open up access to the positions and the interlinking of strands that construct knots. CAD is considered a disembodied or hands-off practice because it disconnects simulation and reality and ignores relational understanding (Sennett, 2008, pp. 42–43). To overcome the shortcoming of virtual knotting, bodily engagement in the process of drawing and/or 3D modeling would be helpful. One available tool in the Mixed Reality Lab was a VR station. The author's hands holding VR controllers to draw in a three-dimensional VR space a scaled-up knot structure imitated a gestural manner to real-world hand-knotting of string. Drawing in VR evoked her hands-on experience and relational understanding of positions of strands in

FIGURE 3.7

(A) Crafting the knot structure in VR, (B) a CAD model of a section of flexible knots, and (C) a print with missing parts and a complete print.

three-dimensional space that she had possessed while hand-knotting. The new experience in gestural crafting in VR augmented the author's understanding of the positions of strands of knots and assisted her in finding a solution for making a CAD model of a section of knots (Fig. 3.7). Working with software is not always a disembodied practice as Sennett argues, as demonstrated by the author's experience of gestural crafting in VR.

Using the 3D model in Fig. 3.7B for 3D printing posed a challenge—the printing nozzle irritated an earlier printed area with a steep angle, causing it to shift from its original position on the support material and resulting in the detachment of the next printed layer. Upon the removal of the support material, initial prints came apart or had a fractured and uneven surface if they stayed intact. It was hypothesized that the printing problem might arise from two factors: the machine and the parameter settings on the printer's software. To test the hypotheses, first, the same 3D model was printed on different machines. The results improved, yet cracked surfaces still occurred. Then, the slicing parameters were set to produce a full grid of support material. A new print was successful, but the support material was too dense to remove. The testing proved that while the machine to some extent affected the printing process, the setting of printing parameters was a more vital factor. The support material was set to spread throughout, strong but reasonably easy to remove. After that, this approach was used to model and print multisectional loose knots which yielded satisfying results (Fig. 3.8).

3.6 Translational craft: when digital meets analog materials experience

The Digital Fabrication Lab obtained powder- and binder-based printing technology for mold making for use with castable materials, such as metal, glass,

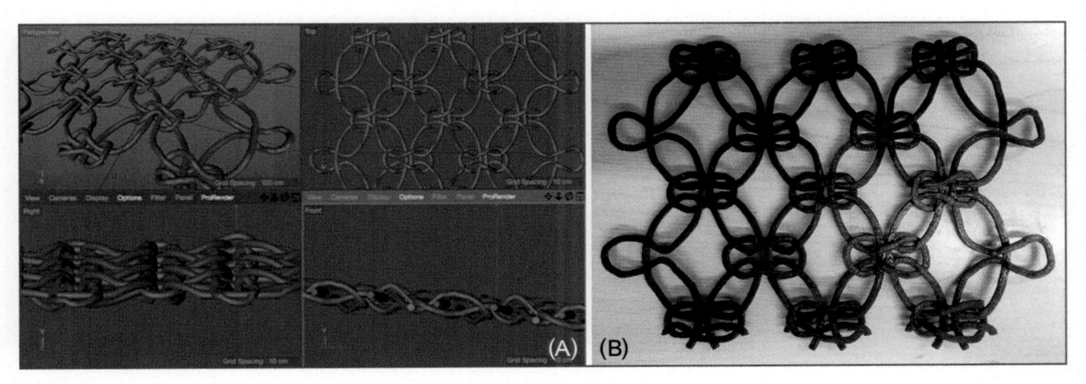

FIGURE 3.8
(A) A CAD model and (B) a print of multisectional loose knots.

and clay slip. The lab custom-made a material recipe workable with the Zcorp 310+ printer by combining three constituents of Hydroperm (i.e., gypsum-based plaster), calcium carbonate, and maltodextrin (i.e., a fine-grained sugar used in the brewing industry) (Robbins et al., 2014, pp. 134–135). While the combination of Hydroperm and calcium carbonate produces good spreadability and relatively high resolution, maltodextrin diffusing through the combined powder creates strong "green strength" (n.d.), that is, mechanical strength to withstand mechanical operations.

At first, the author could not relate her practice with this 3D-printing technology. However, after the accumulation of her digital craft skills the author realized an opportunity offered by this 3D-printing method for giving function to the cup. The author's newly acquired skills were combined with her knowledge of mold making for prototyping and traditional ceramics obtained during her Bachelor degree in Industrial Design to create a CAD model of a two-piece mold for slip-casting a cup with a relief surface of the knot pattern (Fig. 3.9). Based on the 3D model of the knotted cup used previously for 3D printing using PLA filament (Fig. 3.3), the 3D modeling process began with the creation of positive form of the cup and then a 1-in.-thick mold around it.

After being printed and removed from the printer, the mold was depowdered with compressed air, misted with water to increase plasticity, and when completely dry clay slip was cast in (Fig. 3.10). Due to the properties of the gypsum-based material of the 3D-printed mold, the principles of traditional slip-casting clay in a plaster mold could not be followed. For example, the material's higher density required significantly longer time for the cast piece to set, and its brittleness did not allow for the tying of the two-piece mold together, thus turning the mold upside down to drain the excessive slip according to the common practice was impractical. The meeting between the mold made of a nontraditional

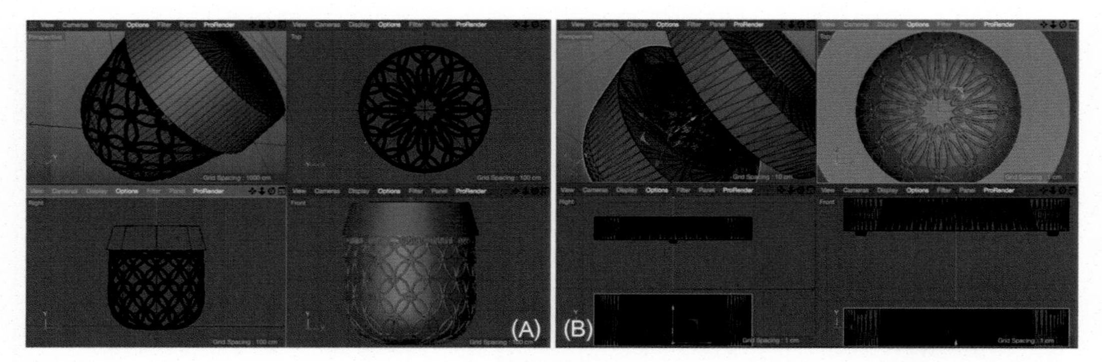

FIGURE 3.9
(A) A CAD model of positive form of the cup which is used to make (B) a CAD model of a 1-in.-thick mold.

FIGURE 3.10
The mold is removed from the printer and the process of using it for slip-casting.

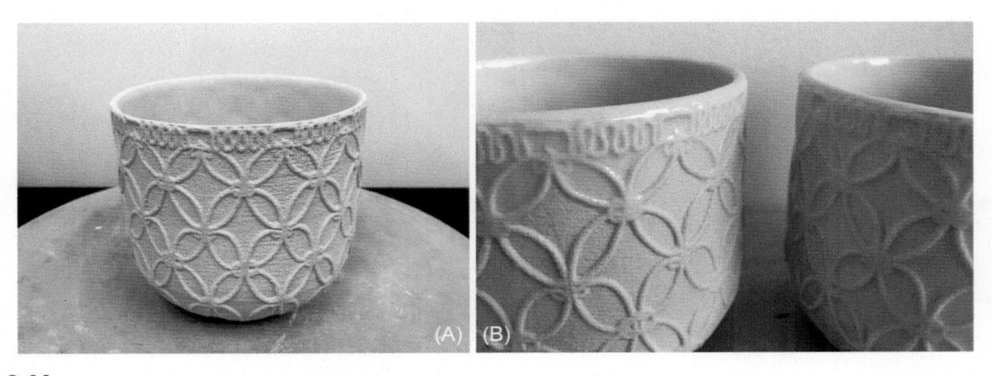

FIGURE 3.11
(A) A cast cup before firing; (B) fired and glazed cast cups, the left being fully glazed inside and partially glazed outside while the right being fully glazed inside and unglazed outside.

material and clay slip which is a traditional material illustrated the agency of a design material as an emergent relationship independent from its fundamental physical qualities. In other words, the relationship between the properties of clay slip and those of the mold's material emerged, manifesting the way they should be used in the casting process in which the clay was continuously shaping and negotiating its ongoing dialog with the mold and the maker. The affordances of the clay slip suggested that it can be cast but it is the agency of the slip or the emergent relationship the slip had with the mold that informed the maker how to deal with this situated casting process in which her knowledge in traditional ceramics needed to be adjusted.

The resulting cast cup revealed an inimitable texture that was the imprint of layers of the powder material, demonstrating the printing process of the mold (Fig. 3.11).

3.7 Discussion: knowing through analog and digital materials experience

Analog and digital materials experience is interconnectedly exercised in the craft practice presented in this chapter. The practice involves action, doing things that are consequential, or what Pickering (2010) calls "performance" or "what agents do, whether human and non-human" (p. 195). Action here is agency, emergent in and unknowable prior to a continuous ongoing performance between humans and nonhumans in a specific context, or in this case, handcrafting through digital means in a lab environment. When the author observed her sense of agency in the real-world experience of working with analog and digital materials in proximity, she felt that she was in a creative collaboration with the materials and tools

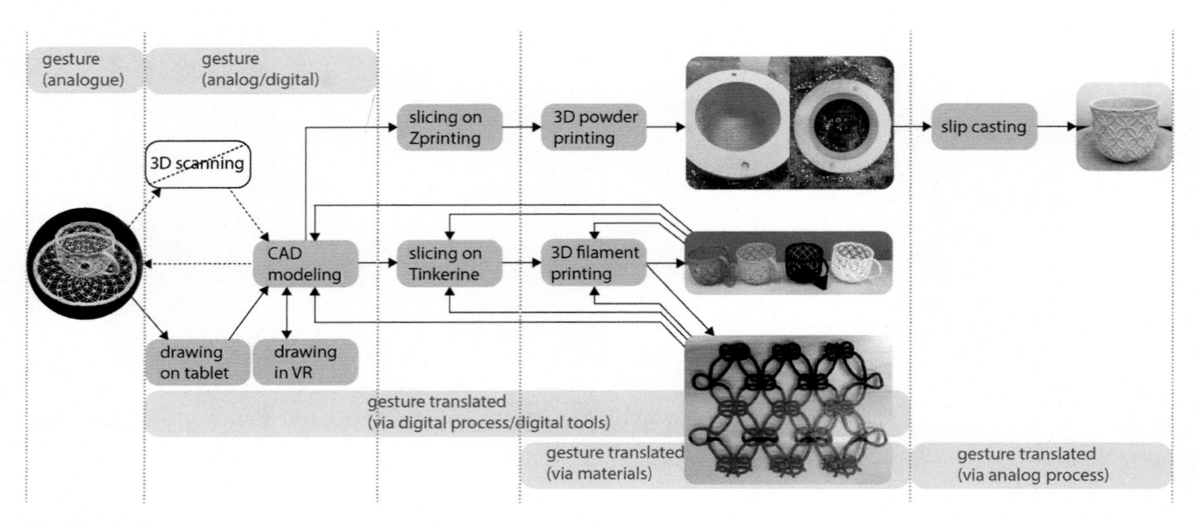

FIGURE 3.12
Work flow of analog and digital materials experience in translation.

that she used, through her gestures in the flow of the activity (Fig. 3.12). Sometimes she acted as a human agent and then became passive when the digital tool (e.g., 3D printer) and/or materials took over the active role—this series of actions went back and forth, and eventually an artifact materialized. Without adopting the materialness or the MDD approaches as such, when dealing with a physical material, such as composite filament, the author's knowledge about these approaches naturally influenced her to let the material take the lead and observed its interaction with the printing process in order to decide on the next experiment. Although the materials and tools have their specific properties and capabilities, their agency or action in a material engagement situation like this craft practice is not fixed but emerges in spatiotemporal relations when the material, the maker, and the tool are interacting and negotiating (Malafouris, 2008), or in a "dance of agency" in Pickering's (2010) term. Materials can thus be understood not as having different states that change according to human action but as performing agents that consequently change the situation. This continuous negotiation between the agents in a situated activity is similar to the notion of "reflective practice" (Schön, 1983) whereby the practitioner reflects on what has happened due to her action, by constantly asking herself where, how, and why it has happened. In this situation, the practitioner can learn from the real-life practice and, with the lesson learned, improve her subsequent practice.

The agency's aspect of unknowability in advance of performance can be looked at from Pye's (1968) "workmanship of risk" (pp. 20–24) or Risatti's (2007) "craftsmanship" (p. 165) that encompasses an uncertain enterprise incorporating material experiment, skillfulness, complexity, and intimacy over time, the results of which is unforeseen. Direct experience with materials, whether physical or digital, can therefore be considered phenomenological investigation. Craftsmanship involves the dialogical/dialectical process that springs from "the formalization of both idea-concept and form-concept in the hand as material is actually being worked via technique" (Risatti, 2007, p. 184). Design, in general, may not involve such a process because the conceptualization stage may be separated from the making stage.

While working in dematerialized settings such as drawing on a Wacom tablet or in the VR space using a goggle and controllers, the author felt the need to intertwine overtly with gesture, the body, and materiality. The esthetic value of the hand was not lost but rather dispersed into a different form. Moving the hand to draw on the tablet or moving the whole body to draw in VR was the act of sense making and that was how the author learned to understand the knot's construction virtually. In this activity, the maker's mind extended itself to the elements of a digital environment experienced through her body.

> [W]e learn our bodies by moving and in moving both create and constitute our movement as a spatiotemporal dynamic. . . . [T]hat movement is the originating ground of our sense-makings . . . [that] we constitute space and time originally in our kinesthetic consciousness of movement. Flux, flow, a streaming present, a stream of thought, consciousness, or subjective life, a style of change—all such descriptive terms are in both a temporal and spatial sense rooted in originary self-movement: they are all primordially present not in the constitution of objects but in our original spontaneity of self-movement, in our original experience and sense of our dynamically moving bodies. (Sheets-Johnstone, 2011, p. 139)

Translation of material and form through the interactive use of analog and digital tools affords new opportunities by means of gestural action and tacit knowledge (Polanyi, 1962). The new experience with the digital tools can be combined with the previously acquired craft knowledge embodied in the maker. The flow in the material translation suggests the emergent material practice, in which an embodied understanding of meaning of the practice is gained by being close to the constantly changing material. Here, meaning is not "the product of representation but the product of a 'conceptual integration' between material conceptual domains" (Malafouris, 2013, p. 90). The use of new tools and materials offers new outcomes and new meanings created through the maker's interaction with materials and tools not only in space but also over time.

Shifting to a more immersive digital tool such as VR to give form and structure of knots, the author was able to repossess her embodied experience of knotting string and create three-dimensional vector-line structures, even though the physical material was absent in the VR space. The embodied knowing in material practice was translated directly to her facility for virtual knotting. Embodied knowledge, according to Merleau-Ponty (1962), is "knowledge in the hands, which is forthcoming only when bodily effort is made, and cannot be formulated in detachment from that effort" (p. 166). Bodily sensorimotor experiences play a fundamental role in the structure of our thinking (Malafouris, 2013, p. 67). Embodied knowledge is sensory and grounded in bodily experience, situating "intellectual and theoretical insights within the realm of the material world" (Ellingson, 2008, pp. 244–245). It is a way of knowing where the mind is inseparable from the body (Lakoff & Johnson, 1999) to gain insights into materials experience.

Craft knowledge can be derived from "intensive and extensive experience ... [that] includes the whole body sensitivity" (Hardy, 2004, p. 181). Through experience, the craftsperson can identify different materials by using her senses and can tell how a certain material will behave according to her manipulation or interaction with it. "Craftsperson, medium, [and] tools ... are of one unified gesture of creativity; their boundedness is less metaphysical than physical" (Hardy, 2004, p. 181). What craftspeople make is "a record of what went on in the studio between maker and materials. Every mark left on an object is a record of a decision made and action taken by the hand that formed it. ... [T]he object represents a subtle reenactment of maker with material" (Fariello, 2004, p. 151). In the case presented in this chapter, the 3D-printed artifacts with their distinctive marks of missing details are physical objects that document not only the digital process and material agency but also the author's decision and action made in the manipulation of the parameter settings of the 3D printer that in turn leave marks on the resulting prints. The author works inductively and integrates material and experiential sensibility into the design process. In so doing, the discovery and generation of new problems can be part of problem-solving processes in prototyping commonly known in design processes. The inductive process has resulted in not only the achievement of objects but also a new way of thinking. The craft practice presented in this chapter is an example of how a craft practitioner-researcher attempts to discover ways of translating handcrafted artifacts into a new form of craft using digital tools. The outcomes show how knowledge is gained through the experience of working in dialog with the analog and digital means and how the intertwined analog and digital materials experience emerges.

Acknowledgments

The author is grateful to the Material Matters Research Center, Emily Carr University of Art + Design, and faculty members there, including Aaron Oussoren, Hélène Day Fraser, Keith Doyle, Logan Mohr, Sean Arden, and Julie York.

References

Adamson, G. (2007). *Thinking through craft*. Berg: Oxford.

Campbell, J. R. (2007). *Digital craft aesthetic: Craft-minded application of electronic tools*. In G. Follett (Ed.), *Proceedings of new craft future voices* (pp. 47–64). Dundee: Duncan of Jordanstone College of Art & Design.

Campbell, J. R. (2016). Foreword. In N. Nimkulrat, F. Kane, & K. Walton (Eds.), *Crafting textiles in the digital age* (pp. xis–xxiii). London: Bloomsbury.

Clark, A., & Chalmers, D. (1998). The extended mind. *Analysis, 58*(1), 7–19.

Cross, N. (2008). *Engineering design methods: Strategies for product design* (4th ed.). Chichester: John Wiley & Sons.

Desmet, P., Hekkert, P., & Schifferstein, R. (2011). Introduction. In P. Desmet, & R. Schifferstein (Eds.), *From floating wheelchairs to mobile car parks: A collection of 35 experience-driven design projects* (pp. 4–12). Den Haag: Eleven.

Dewey, J. (1934). *Art as experience*. New York: Perigee.

Dormer, P. (1997). Craft and the turning test for practical thinking. In P. Dormer (Ed.), *The culture of craft* (pp. 137–157). Manchester: Manchester University Press.

Ellingson, L. L. (2008). Embodied knowledge. In L. M. Given (Ed.), *The Sage encyclopedia of qualitative research methods* (pp. 244–245). Thousand Oaks, CA: Sage.

Fariello, M. A. (2004). "Reading" the language of objects. In M. A. Fariello, & P. Owen (Eds.), *Objects meaning: New perspectives on art and craft* (pp. 148–173). Lanham, MD: The Scarecrow Press.

Frayling, C. (1993). Research in art and design. *Royal College of Art Research Papers, 1*(1), 1–5.

Gell, A. (1998). *Art and agency: An anthropological theory*. Oxford: Clarendon Press.

Giaccardi, E., & Karana, E. (2015). Foundations of materials experience: An approach for HCI. In *Proceedings of the 33rd SIGCHI conference on human factors in computing systems* (pp. 2447–2456). New York: ACM.

Gray, C., & Burnett, G. (2009). Making sense: An exploration of ways of knowing generated through practice and reflection in craft. In L. K. Kaukinen (Ed.), *Proceedings of crafticulation and education conference* (pp. 44–51). Helsinki: NordFo.

green strength. (n.d.). In *McGraw-Hill Dictionary of Scientific & Technical Terms, 6E*. (2003). <https://encyclopedia2.thefreedictionary.com/green + strength> Accessed 01.06.20.

Hardy, M. (2004). Feminism, crafts, and knowledge. In M. A. Fariello, & P. Owen (Eds.), *Objects meaning: New perspectives on art and craft* (pp. 176–195). Lanham, MD: The Scarecrow Press.

Heidegger, M. (1962). *Being and time* (J. Macquarrie & E. Robinson, Trans.). London: Basil Blackwell (Original work published 1927).

Karana, E., Barati, B., Rognoli, V., & Zeeuw van der Laan, A. (2015). Material driven design (MDD): A method to design for material experiences. *International Journal of Design, 9*(2), 35–54.

Karana, E., Hekkert, P., & Kandachar, P. (2008). Materials experience: Descriptive categories in material appraisals. In *Proceedings of the conference on tools and methods in competitive engineering* (pp. 399–412). Delft: Delft University of Technology.

Karana, K., Pedgley, O., & Rognoli, V. (2014). *Materials experience: Fundamentals of materials and design*. Amsterdam: Elsevier.

Karana, K., Pedgley, O., & Rognoli, V. (2015). On materials experience. *Design Issues, 31*(3), 16–27.

Knappett, C. (2005). *Thinking through material culture: An interdisciplinary perspective*. The University of Pennsylvania Press: Philadelphia, PA.

Lakoff, G., & Johnson, M. (1999). *Philosophy in the flesh: The embodied mind and its challenge to western thought*. New York: HarperCollins.

Malafouris, L. (2008). At the potter's wheel: an argument for material agency. In C. Knappett, & L. Malafouris (Eds.), *Material agency: Towards a non-anthropocentric approach* (pp. 19–36). New York: Springer.

Malafouris, L. (2013). *How things shape the mind: A theory of material engagement*. Cambridge, MA: MIT Press.

McCullough, M. (1996). *Abstracting craft: The practiced digital hand*. Cambridge, MA: MIT Press.

Merleau-Ponty, M. (1962). *Phenomenology of perception* (C. Smith, Trans.). London: Routledge Classics (Original work published 1945).

Niedderer, K., & Townsend, K. (2014). Designing craft research: Joining emotion and knowledge. *The Design Journal, 17*(4), 624–647. Available from https://doi.org/10.2752/175630614X14056185480221.

Nimkulrat, N. (2009). *Paperness: Expressive material in textile art from an artist's viewpoint*. Helsinki: University of Art and Design Helsinki.

Nimkulrat, N. (2010). Material inspiration: From practice-led research to craft art education. *Craft Research Journal, 1*(1), 63–84.

Nimkulrat, N. (2012). Hands-on intellect: Integrating craft practice into design research. *International Journal of Design, 6*(3), 1–14.

Pickering, A. (1995). *The mangle of practice: Time, agency, and science*. Chicago, IL: The University of Chicago Press.

Pickering, A. (2010). Material culture and the dance of agency. In D. Hicks, & M. C. Beaudry (Eds.), *The Oxford handbook of material culture studies* (pp. 191–208). Oxford: Oxford University Press.

Pickering, A. (2013). Being in an environment: A performative perspective. *Natures Sciences Sociétés, 21*, 77–83. Available from https://doi.org/10.1051/nss/2013067.

Polanyi, M. (1962). *Personal knowledge*. London: Routledge.

Pye, D. (1968). *The nature and art of workmanship*. Cambridge: Cambridge University Press.

Risatti, H. (2007). *A theory of craft: Function and aesthetic expression*. Chapel Hill, NC: The University of North Carolina Press.

Robb, J. (2010). Beyond agency. *World Archaeology, 42*(4), 493–520. Available from https://doi.org/10.1080/00438243.2010.520856.

Robbins, P., Doyle, K., & Day-Fraser, H. (2014). Hybridizing emergent digital methodologies across legacy creation ecosystems. In *All Makers Now? 2014 Journal*. Falmouth: Falmouth University. <http://www.autonomatic.org.uk/allmakersnow/wp-content/uploads/2015/07/AMN2014_Robbins_et_al.pdf> Accessed 01.06.20.

Schön, D. A. (1983). *The reflective practitioner: How professionals think in action*. New York: Basic Books.

Sennett, R. (2008). *The craftsman*. New Haven, CT: Yale University Press.

Sheets-Johnstone, M. (2011). *The primacy of movement: Expanded second edition*. Amsterdam: John Benjamins.

Shillito, A. M. (2013). *Digital crafts: Industrial technologies for applied artists and designer makers*. London: Bloomsbury.

Zoran, A., & Buechley, L. (2013). Hybrid reassemblage: An exploration of craft, digital fabrication and artifact uniqueness. Leonardo 46(1), 4–10. Available from https://doi.org/10.1162/LEON_a_00477.

CHAPTER 4

Digital crafting: a new frontier for material design

Manuel Kretzer
Materiability Research Group, Dessau Department of Design Anhalt, University of Applied Sciences, Dessau, Germany

> Digital machines and productive technologies in general allow for the production of an industrial continuum. From the mold we move toward modulation. We no longer apply a preset form on inert matter, but layout the parameters of a surface of variable curvature. A milling machine that is commanded numerically does not regulate itself according to the build of the machine; it rather describes the variable curvature of a surface of possibility. The image-machine organization is reversed: the design of the object is no longer subordinated to mechanical geometry; it is the machine that is directly integrated into the technology of a synthesized image.
>
> **Bernard Cache, *Earth moves: The furnishing of territories* (1995)**

4.1 Introduction

The term "Digital Crafting" has its origins in architecture and describes the regained capacity of the architect, empowered through novel design and manufacturing methods, to have a direct and personal impact on craftsmanship or even more so to become self-active in artisanal processes. Just as in the early days of architecture, when the master builder was still involved in all areas of building—from initial planning and ideation until its physical completion—these new technologies allow a return to the exploration and testing of experimental design methods and the direct exchange with a multitude of different materials. Designing for and through digital production techniques thus shifts the focus from formal design representations toward the physically realized. As such, material and tectonic thinking are reintroduced as the very base and starting point of the design approach. Due to this shift of focus a new type of design becomes possible with a formerly unknown degree of complexity—both on a formal and on a functional level (Sachs, 2012).

Materials Experience 2. DOI: https://doi.org/10.1016/B978-0-12-819244-3.00003-X

The French architect and theorist Bernard Cache, an early adopter of what he calls Computer-Assisted Conception and Fabrication systems, imagines that such thinking, where objects are not any more drawn but calculated, will enable the creation of highly complex forms that would be difficult or impossible to draw in a conventional way. Equally interesting he proclaims that such systems will incept a "nonstandard mode of production" where items from the same series can vary in size and shape by simply altering certain fabrication parameters. Since thus neither an object's function nor its materiality is linked to a defined form but can vary constantly, the digital representation "takes precedence over the object" (Cache et al., 1995, p. 97). The core representation of an object is therefore not any longer its image but a model of simulation, an instance of numerical manipulation.

Digital crafting, therefore, refers to much more than just the use of digital technologies to produce objects. It is not only about the enhancement of the craftsman's range of tools through devices such as Computer numerical control (CNC) milling machines, three-dimensional (3D) printers, or laser cutters, but rather about the unprecedented possibilities that these technologies reveal. For the laser cutter, it makes no difference whether the shape being cut is a rectangle or a high-detail fractal, as long as the required data are provided. The CNC milling machine does not care if each of the objects it produces is unique or if it has to "work" overtime. The 3D printer offers a 3D void of theoretically unlimited geometric opportunities. For the designer, however, this new freedom means not only becoming aware of each of these possibilities and likewise limitations, but above all understanding and using them adequately.

But what does adequate mean when there are almost no restrictions? In order to figure out what these processes and machines are best suited for, one should perhaps follow the approach of Louis Kahn and very carefully try to identify what our tools and materials really want.

> You say to a brick, 'What do you want, brick?' And brick says to you, 'I like an arch.' And you say to brick, 'Look, I want one, too, but arches are expensive and I can use a concrete lintel.' And then you say: 'What do you think of that, brick?' Brick says: 'I like an arch.' (Kahn & Twombly, 2003, p. 270)

To understand what a 3D printer's printing process and material thus are best suited for and hence discover the true potential but also limitations of digital crafting, designers and architects alike must acquire new skills that go beyond what is presently being taught and practiced with conventional methods. Digital crafting therefore not only describes physical results but even more so a concept, an idea, or a method, which—driven by current digital design and production techniques—implies a new way of dealing with, and a seamless connection of, form, construction and materiality. It is not so much a matter of the classical design of products and their manufacture by means of computer-controlled machines, but rather of acquiring practical and theoretical skills in dealing with digital processes for generating form and function.

To better comprehend the larger meaning of this concept it might help to have a brief look at the history and evolution of digital design and manufacture.

4.2 A brief history of digital design and fabrication

The beginnings of numerical control (NC) are generally credited to John T. Parsons, who in 1949 was asked by the US Air Force to develop a cost-effective, automated method for the production of helicopter rotor blades. Parsons teamed up with the Servomechanisms Laboratory at the Massachusetts Institute of Technology, who back then was on the global lead of mechanical computing and feedback systems, and together they created the computer-controlled "Card-a-Matic Milling Machine" for three-axis contour milling. Even though the device was relatively difficult to set up and only able to perform a small number of operations it marked a revolution in automated manufacture. The invention reduced labor, downtime, and waste material while increasing productivity, precision, and versatility through one single tool (Caneparo, 2016).

During the 1960s computing became more publicly recognized and a new generation of cyberneticists began wondering about design and its parallels to cybernetic systems (Spiller, 2009). In 1963 Ivan Sutherland published Sketchpad, a pioneering effort in human-computer interaction, often credited as the very first computer-aided design (CAD) tool. Sutherland himself described Sketchpad as a system that, "by eliminating statements (except for legends) in favor of line drawings, opens up a new area of man-machine communication," which in the past "has been slowed down by the need to reduce all communication to written statements that can be typed." Sutherland saw the benefits of his innovation especially in the making of drawings where motion, analysis, high accuracy, or repetition was crucial, yet he conceded that "it is only worthwhile to make drawings on the computer if you get something more out of the drawing than just a drawing" (Sutherland, 1998, p. 17, 99, 110).

Graphical CAD tools were further enhanced in the 1970s and became accessible to designers and architects. At the same time NC systems were improved and computer-aided manufacturing (CAM) was developed. Its direct link to CAD helped, in that it was rapidly accepted by various industries for the creation of complex products such as ships or automobiles. While in the beginning the technology was only practical for projects with large volumes developed by sizeable companies, the continuous increase in computing power and the reduction in costs ultimately led to the propagation of CAD/CAM and CNC in the fields of industrial design (Computer-Aided Industrial Design) and manufacturing (Corser, 2012).

In parallel, additive manufacturing, also known as 3D printing, appeared on the market in the 1980s and quickly evolved through numerous companies that invested in its development (Gershenfeld, 2012). Around the same time, industrial robots gained momentum due to their great versatility. Since the "end-effectors, attached and controlled by these arms, are as diverse as the materials they can process," much of today's research into digital fabrication still focuses on the use of robotic arms (Beorkrem, 2013).

Neil Gershenfeld, head of MIT's Center for Bits and Atoms, claims that continuous evolution of these fabrication technologies will inevitably result in a "new digital revolution," empowering people "to design and produce tangible objects on demand, wherever and whenever they need them." Encouraged by the global success and rapid growth of Fab Labs, he believes that the true power of such technologies "is not technical; it is social," enabled through their largely unregulated but well-connected nature. Gershenfeld proposes that the ultimate capabilities of what he refers to as "assemblers"—the future offspring of 3D printers—will be "to create complete functional systems in a single process" without any waste or left-overs (Gershenfeld, 2012).

4.3 The digital continuum

Today, digital design and fabrication technologies have arrived at almost every design and architecture school and many creative companies have established their own computation units and in-house fabrication and prototyping facilities. To eventually reach what Branko Kolarevic, Dean of the College of Architecture and Design at New Jersey Institute of Technology, calls "a digital continuum from design to production" (Kolarevic, 2003, p. 7) the focus in education should be on learning methods of digital form genesis, such as the examination of parametric processes, which no longer describe a shape in concrete terms, but define it through the correlation of variables, parameters, and explicit rules. A parametrically designed object is not limited to a certain manifestation, but rather acts within a dynamic space of action and can be realized in any form between previously defined boundaries (Böhm, 2001).

For example, the following program, written in the Java-based programming language Processing, draws a regular polygon defined only by the two variables "radius" and "corners" (Fig. 4.1):

```
int radius = 10;
int corners = 3;
float angle = TWO_PI / corners;
beginShape();
```

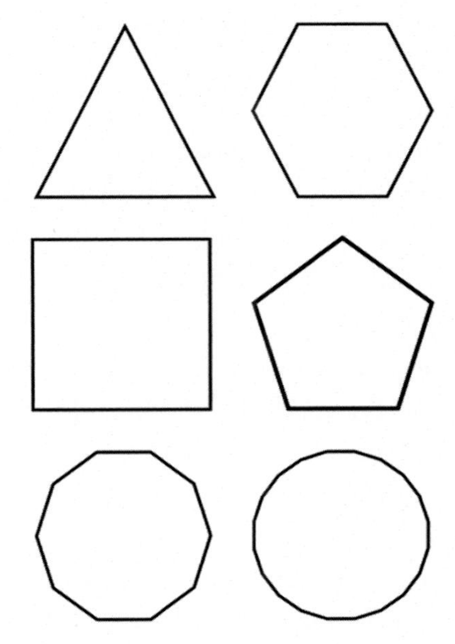

FIGURE 4.1
Different polygons created from a single basic algorithm.

```
for (float a = 0; a < TWO_PI; a + = angle) {
  float sx = width/2 + cos(a) * radius;
  float sy = height / 2 + sin(a) * radius;
  vertex (sx, sy);
}
endShape(CLOSE);
```

Simply by changing the value of the two variables, any polygon can be drawn with any radius and any number of corners. Although this is a relatively banal example, it illustrates the difference to the classic use of CAD systems, which are merely a kind of electronic drawing board.

This approach becomes even more interesting when the computer no longer exclusively executes what is imagined (or programmed) by a human designer but gets integrated into the design process in a "creative" way. This technique, often defined as "generative" or "procedural" drawing or design, not only enables the creation of geometries and shapes with increased complexity and detail, but above all produces results that are not any longer directly predictable and controllable. For example, using a script similar to the previous, one can now generate an infinite variety of similar but yet unique forms (Fig. 4.2):

```
float[] x = new float[4];
float[] y = new float[4];
for (int a = 0; a < 4; a++) {
  x[a] += random(-20, 20);
  y[a] += random(-20, 20);
}
beginShape();
curveVertex(x[3] + width/2, y[3] + height/2);
for (int a = 0; a < 4; a++) {
  curveVertex(x[a] + width/2, y[a] + height/2);
}
curveVertex(x[0] + width/2, y[0] + height/2);
curveVertex(x[1] + width/2, y[1] + height/2);
endShape();
```

One of the many advantages of this type of design approach is that by automatically generating endless variations, the designer's focus can shift away from manual drawing to other, possibly more exciting or important aspects of the design. The transfer of abstract thought constructs into drawn geometry gives way to an active selection process based on certain preferences. The result, which is only vaguely known at the beginning, is only discovered and continuously developed as the process progresses. The result-oriented design thus becomes the process-oriented design.

4.4 Individual production

Another innovation resulting from parametric design concepts and digital fabrication methods is the so-called "Mass Customization." Mass customization describes the idea of adapting serially manufactured products to individual customer requirements without increasing costs or manufacturing effort. The pioneer of digital production processes, Prof. Emeritus Yoram Koren, goes even one step further when describing the factory of the future as a production facility that is directly connected to the customer, producing individually tailored products. In order to make this possible, current manufacturing procedures must be adapted from mass to individual production and interface tools need to be developed that allow the layperson to enter into a direct exchange with the manufacturing company (Clifton, 2018). Although the ideas of mass customization and individual

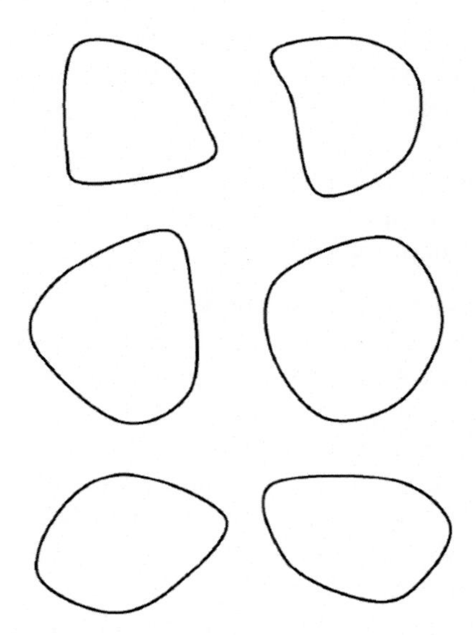

FIGURE 4.2
Multiple organic forms created from a single simple script.

production have existed since the early 1990s, they to date have not been really implemented, or at least only to a very limited extent (Modrak, 2017). The reasons for this are insufficiently mature digital technologies that cannot yet compete with the established manufacturing cycles, industrial infrastructures that favor existing systems over novel, more experimental approaches, and finally a lack of understanding within the creative community for the potentials of such approaches and a lack of educational methods for their transfer.

However, this could soon change fundamentally. In his book "Makers: The New Industrial Revolution," Chris Anderson, former editor-in-chief of the magazine "Wired," describes the productive efficiency of the digital world and its impact on society. Anderson argues that with the advent of consumer-oriented digital fabrication tools such as desktop 3D printers, as well as instant access to cloud-based manufacturing facilities, information that was previously largely confined to a "digital existence" is now finding its way back into the physical world. For him, this transformation of bits into atoms is the logical development from mass production to mass customization and thus represents the "third industrial revolution" (Anderson, 2014, p. 17).

4.5 Toward a new esthetics

This change is also noticeable in a new kind of esthetic formalization that goes hand in hand with digital processes. Things that were previously undrawable and perhaps even unimaginable can now not only be visualized, but also materialized instantaneously anywhere and anytime using 3D printing processes. Even if—in terms of large-volume production—the fabrication costs of these products cannot yet fully compete with more traditional methods, a clear tendency becomes noticeable that explicitly distances itself from the minimalism and reductionism of the previous century. Intricacy, detail, and ornament return.

It remains exciting to see how the role of the designer will change and develop. While in architecture, styles have often alternated with sometimes more and sometimes less ornamental decoration, the history of design is much younger (Kretzer, 2017, p. 46). Design has its roots in the modern forms of mass production. Although Greek coins or Roman amphorae were already produced in large quantities, it was basically craftsmen who were responsible for both their design and their manufacture. It was only with the industrialization and mechanization of production methods during the 18th century that this tradition was broken, and the role of the designer was created. Designers were given the task of creating practicable solutions for industrial production processes that required considerable investment but could produce large quantities of identical objects at a low cost. Detail, individuality, and uniqueness gave way to efficiency, simplicity, and functionality.

Digital crafting undermines this ideal of modernity, the concept of identical reproduction for the masses. Because everything that is digital remains variable and changeable and thus stands in stark contrast to the idea of standardization and serial production. The theorist Mario Carpo calls this the first phase of digitalization, which inverts the industrial revolution and thus the necessity of mass manufacture. But Carpo goes even further and argues that now, enabled by artificial intelligence, we will experience a second phase: digitalization on an intellectual level. Since for him the first step means a return to traditional production methods and the reunification of the designer and the craftsman into one and the same person, he now believes that we will also find our way back ideologically, politically, and socially to prescientific, intuitive, even medieval systems and structures (Sabin & Carpo, 2017).

Although the hope is that—as the term Digital Crafting suggests—the importance of the topic will remain primarily focused on its physical aspects and that an evolution in design and production methods will not be accompanied by social regression, its relevance to the field of design and architecture is undeniable.

To understand these new approaches less as a replacement of existing systems and processes, but rather as a necessary addition, the designer's task will be to navigate within this spectrum of possibilities and select the most appropriate solution according to context, client, and demand. In this context—to mediate the meaning and possibilities of digital crafting—students should be encouraged to focus on the new opportunities that arise from the unification of design and fabrication. This includes the examination of different manufacturing techniques and new technologies, but most of all the experimental study of materials and their confluence with digital methods. The playful exploration of such processes, the growing confidence in their usage, and the autonomous navigation in related communities and networks are essential skills that today's students need to develop.

4.6 Digital crafting in educational practice

The following are a selection of projects, realized as part of an ongoing collaboration with the German car manufacturer AUDI, where digital crafting as a method to design and fabricate form, was used to investigate scenarios of future mobility. The focus was particularly on subjects like autonomous driving and car-sharing platforms, which will increase the need for personal customization and adaptation of the vehicle's interior.

The first project, "Concept Breathe," was realized within 3 months with a group of 10 second and fourth semester Bachelor students from the Braunschweig University of Art in 2017. It aimed at rethinking the image of the classical car seat. The project began with an intense ideation phase, looking particularly into patterns and structures that can be found in nature. The team then studied how these inspirations can be described using computer scripts and algorithms in order to lead to a design language that deviates from conventional forms and to develop a structurally sound, lightweight system to house an array of active components. After computationally designing numerous organic forms, 3D printing was chosen as the most viable method for the 1:1 realization of a prototype mock-up. Supported by BigRep, a Berlin-based producer of large-scale 3D printers, the design was printed from Pro-HT Plastic, a biodegradable polymer with high-strength properties. In addition to the 3D-printed structure, the seat contained 38 custom-made, active silicon cushions that were incorporated into its surface to dynamically alter its visual and haptic properties. The components were intended to increase the seat's ability to respond to changing driving conditions and particularly to enhance the user's identification with the animate object through motions of breathing. Finally, a number of customized cushions from a high-performance fabric were added in five separate areas to guarantee the necessary comfort and stability (Fig. 4.3).

Building upon the experience from the first project, a second collaboration took place in 2018, this time at the Dessau Department of Design, Anhalt

University of Applied Sciences, and with a focus on the middle console. The students worked in groups of four, each developing their own specific scenario and design approach. Requirements, which each group had to address, were functional aspects (e.g., tray or cup holder options), formal expressions (using parametric/generative processes), technological production (applying digital fabrication techniques), sustainability (e.g., materiality and recycling), interaction with and adaptation of the system, and finally embedding the object into a larger narrative context. The project "Alcyon" proposed a bio-inspired center console produced by 3D printing, using a biodegradable filament made from algae. The structure was further planted with reindeer lichen—an organism consisting of algae and fungi, living in symbiosis without the need for human intervention. Lichens were used as bioindicators to demonstrate habitable air quality while the algae provided oxygen through photosynthesis, impacting and improving the indoor climate. The project "Okura" focused on the identification of the driver with the vehicle. While today a driver is an active part of a large system, the shift will be more toward passive scenarios where the user will become a mere passenger. To compensate for loss of control, alienation, and a lack of security, the connection between driver and machine must be improved. Okura represented an

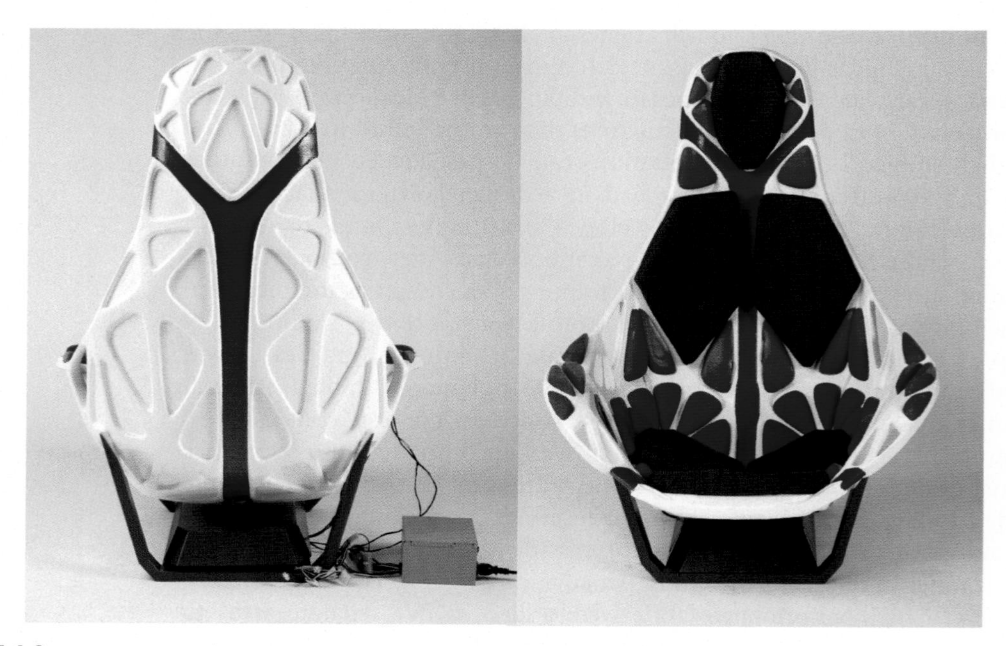

FIGURE 4.3
Concept Breathe was a 3D-printed car seat, which incorporated active components to dynamically alter its surface to suit varying driving situations.

interactive surface, which informed, adapted, and reacted. The dynamic scales of the surface moved according to behavioral patterns, translating (e)motions into the interior and thus establishing trust in autonomous driving and artificial intelligence. Moreover, the surface could change its color from black to white in response to a change in temperature, which improved the interior climate (Fig. 4.4).

The most recent collaboration with AUDI, which took place in 2019, was called "Adaptive City Car" and was a joint course between the Dessau Department of Design and the Dessau Institute of Architecture. Focusing on the future of urban mobility, which is facing tremendous challenges such as increasing traffic, air pollution, and a lack of car-free zones, the goal of the 3-month course was to build a 1:1 model of an autonomous shared car, since most of today's car-sharing alternatives are missing a number of important factors in comparison to owning a private vehicle. The Adaptive City Car project aimed at radically breaking with current industry standards and offering a hypothetical solution for a more sustainable urban vehicle that would provide better air quality, room for different activities during driving, and most of all the feeling of occupying a personal space. The organic design of the car was based on generative principles that tried to provide an alternative to the artificial and standardized appearance of ordinary vehicles. In addition to a structure that seemed naturally evolved, certain areas of the shell were designed to encourage the growth of moss and lichen to filter air pollutants entering the car's interior. The asymmetric shape of the vehicle allowed for more internal space but also the integration of large skylights, made from opacity-changing smart glass, that directed the visual focus of passengers from unpleasant traffic situations to the environment. Sensors that responded to the presence of individuals controlled integrated light and

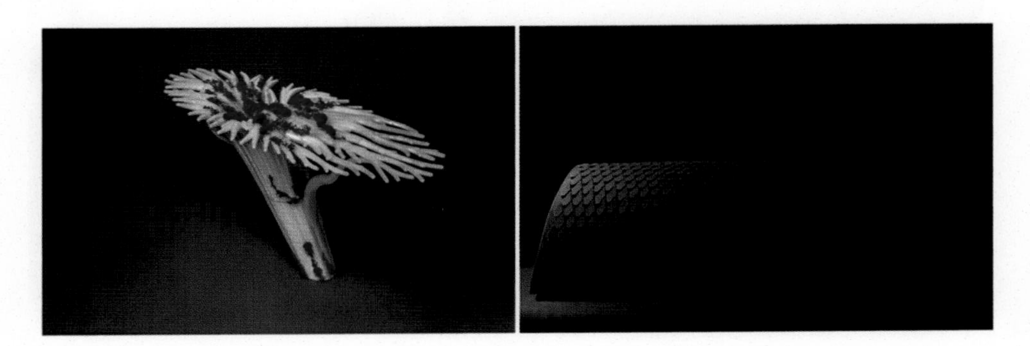

FIGURE 4.4

Left: Alcyon was a 3D-printed center console based on an algorithm that simulates coral growth. The structure was covered with lichen intended to produce oxygen and improve the indoor climate. Right: Okura was a surface made of 1200 dynamic scales that could move and change their color in order to increase the passenger's identification with the vehicle.

audio feedback, in order to increase the passengers' identification with the self-driving vehicle. The 1:1 model was robotically produced from 92 bespoke components that were hand-coated with up to three layers of polystyrene adhesive, sanded, and then assembled together. Once all units were in place the completed vehicle was painted and the adaptive technologies and organic materials were installed (Fig. 4.5).

Digital crafting as a design tool and method allows projects such as those described in this chapter to depart from traditional forms and thinking and instead explore new territories of digital shape genesis, (automated) fabrication, and experimental materiality. Likewise, certain aspects of the design process have lost their importance while others have become more prominent. Especially the creation of drawings, diagrams, and imagery, usually used to communicate the design intent to related fabricators and craftsmen, is being replaced through digital code, readable by the involved machines. However, to reach Kolarevic's digital continuum still many things have to change and evolve. Most of all, there needs to be a shift in thinking and education, raising awareness of the unprecedented opportunities these technologies offer. At the same time the involved tools need further improvement and development in respect to specific design demands and economic viability. And lastly, software and programs need to be simplified, since one of the largest obstacles for their larger impact are their complex interfaces. However, in regard to the current situation, in the middle of a global pandemic with a society gradually transforming into a digital age, it is nowadays more important than ever to engage in new concepts and developments and explore the opportunities that lie within.

FIGURE 4.5

The Adaptive City Car project was a full-scale 1:1 mock-up of a vision for the future of urban mobility. The digitally designed and robotically produced vehicle sensed human presence and responded by changing the opacity of its windows and through light and sound. Porous cork elements penetrated the sinuous surface and encouraged the growth of moss and lichen to filter air pollutants entering the car's interior.

Acknowledgments

Concept Breathe (2017)—Braunschweig University of Art and AUDI AG, I/EK-S1, Development/Innovation. Students: Moritz Boos, Maximilian Dauscha, Leon Ehmke, Lydia Jasmin Hempel, Dong-Kwon Lee, Tim Daniel Ingo Lüders, Vanessa Paladino, Benedikt Schaudinn, Sebastian Spiegler. Supervision: Prof. Manuel Kretzer, Mike Herbig (AUDI AG). Support: csi Mission Findus, BigRep, Moritz Begle, ETH Zürich.

Alcyon and Okura (2018)—Dessau Department of Design, Anhalt University of Applied Sciences. Students: Martin Naumann, Luise Eva Maria Oppelt, Hang Li, Toni Pasternak, Anian Stoib, Eric Dwilling, Caren Flohr, Vivien Gärtner, Francesco Langer. Supervision: Prof. Manuel Kretzer, Mike Herbig (AUDI AG). Support: csi Mission Findus.

Adaptive City Car (2019)—Dessau Department of Design and Dessau Institute of Architecture, Anhalt University of Applied Sciences. Students: Saeed Abdwin, Niloufar Rahimi, Aleksander Mastalski, Neady Oduor, Marina Osmolovska, Ashish Varshith, Jan Boetker, Otto Glöckner, Nate Herndon, Dominique Lohaus, Marie Isabell Pietsch, Katja Rasbasch, Fu Yi Ser, Lam Sa Kiu, Anian Till Stoib, Laura Woodrow. Tutors: Adib Khaeez, Valmir Kastrati, Shazwan Mazlan, Manuel Lukas. Supervision:

Sina Mostafavi, Prof. Manuel Kretzer. AUDI Support: Mike Herbig. CG Artist: Mohammed Mansour. Robotic Support: Carl Buchmann. Video: Yoshua Wilm, Esteban Amon. Additional Support: Vanessa Rüpprich, Anton Roppeld, Mo Sayed Ahmad, Yulia Surova, Melissa-Kim Petrasch, Ludwig Epple, Jessica Bösherz, Vanessa Busch, Niklas Menzel.

References

Anderson, C. (2014). *Makers: The new industrial revolution* (pp. 17–18). New York: Crown Business.

Beorkrem, C. (2013). *Material strategies in digital fabrication* (p. 10) New York: Routledge.

Böhm, F. (2001). Parametrisches entwerfen. In *Archplus 158 Houses on Demand* (p. 73).

Cache, B., Wilcox, J., & Cache, B. (1995). *Earth moves: The furnishing of territories* (pp. 96–97). Cambridge, MA: The MIT Press.

Caneparo, L. (2016). *Digital fabrication in architecture, engineering and construction* (p. 55) Dordrecht: Springer.

Clifton, L. (2018). *Die fabrik der zukunft*. <https://new.siemens.com/global/de/unternehmen/stories/industrie/the-factory-of-the-future.html> Accessed 26.09.20.

Corser, R. (2012). *Fabricating architecture: Selected readings in digital design and manufacturing* (p. 13) New York: Princeton Architectural Press.

Gershenfeld, N. (2012). How to make almost anything: The digital fabrication revolution. *Foreign Affairs*, *91*(6), 45.

Kahn, L. I., & Twombly, R. C. (2003). *Louis Kahn: Essential texts* (p. 270) New York: W.W. Norton.

Kolarevic, B. (2003). *Architecture in the digital age: Design and manufacturing* (p. 7) New York: Taylor & Francis.

Kretzer, M. (2017). *Information materials smart materials for adaptive architecture* (p. 46) Cham: Springer International Publishing.

Modrak, V. (2017). *Mass customized manufacturing: Theoretical concepts and practical approaches* (p. 23) Boca Raton: CRC Press.

Sabin, J., & Carpo, M. (2017). Q&A1. In A. Menges, B. Sheil, R. Glynn, & M. Skavara (Eds.), *Fabricate: Rethinking design and construction* (p. 157). London: UCL Press.

Sachs, H. (2012). Digitales werken. Über die einflüsse und entwicklung der digital produzierten, modularen ausstellungsarchitektur. In *Architekturteilchen. Modulares Bauen im Digitalen Zeitalter* (p. 13) Köln: MAKK.

Spiller, N. (2009). *Digital architecture now: A global survey of emerging talent* (p. 10) New York: Thames & Hudson.

Sutherland, I. E. (1998). Sketchpad—A man-machine graphical communication system. *Seminal Graphics*, 391–408. Available from https://doi.org/10.1145/280811.281031.

Surface texture as a designed material-product attribute

Bahar Şener, Owain Pedgley
Department of Industrial Design, Middle East Technical University, Ankara, Turkey

5.1 Introduction

The research in this chapter comes from the initial phase of the *textureface* project—a portmanteau of "textured surface"—which sets out to investigate and ideate the role of texture as a surface characteristic of products. The end goal of the *textureface* project is to develop principles for texture designs that can demonstrably influence action tendency (Sonneveld & Schifferstein, 2008)—that is, tendency to touch a product (tactual attraction) or tendency to avoid touching (tactual aversion). It is well known that tactility is good for us: that our sense of well-being can be increased through diverse touch experiences and that aiming for "pleasant tactility" within a product (Sonneveld & Schifferstein, 2008) can be a route to satisfying the hedonic element of Hassenzahl's (2003) functional-hedonic user needs analysis. However, although texture provides a means to define product personality, add interest, and increase perceived value, the particular interest in surface texture for this chapter, as will become apparent, is not esthetic expression but rather the less investigated perspective of functionality.

The chapter digs into the subject of how a product's function can be enhanced or delivered through surface textures, and thus how surface texture can be a contributor to the functional element of Hassenzahl's user needs analysis. Existing literature gives only hints about what may be termed as functional texture. A common point is the improvement of grip during the user–product interaction (Özcan, 2008; Zuo et al., 2016; Zuo, 2010). Texture-based wayfinding systems within the floors and walls of buildings, especially for visually impaired people, are another example (Herssens & Heylighen, 2010). Researchers have also examined biomimetic approaches to functional texture, for example, to design product surfaces with enhanced adhesion or hydrophobia, associated with the surfaces of certain biological species (Bhushan & Sayer, 2007; Fratzl, 2007). McCardle (2015) abstracted skin textures of aquatic animals to create 3D-printed surfaces on products required to possess low surface friction while operating in an immersed liquid environment.

Materials Experience 2. DOI: https://doi.org/10.1016/B978-0-12-819244-3.00021-1

While all these examples are interesting, they are specialized, dispersed, and inaccessible to designers. Relatively few research studies on texture and tactility have had intention to directly feed into designers' decision-making processes. This is despite the fact that information on the tactile characteristics of materials is highly requested by professional designers (Karana et al., 2014). Research for tactility within multisensorial frameworks is, however, gaining prominence (Bakker et al., 2015; Spence & Gallace, 2011), with one new design tool—the "experience map" specifically mentioning surface texture in supporting the overall UX vision for a product (Camere et al., 2018). Within the product design field, such research is driven by a motivation to better understand how tactility can mediate the user–product interaction and resultant experiences, especially over relatively long periods of product ownership and use (Fenko et al., 2010).

One of the initial aims for the *textureface* project was to conduct a wide-ranging and systematic analysis of functional texture, the outcomes of which could be used to provide designers with strategies and practical advice. To achieve this, two principal research objectives were set. The first objective was to clearly define and describe texture from a product design perspective. To do this, a characterization of surface texture was carried out, analyzing four key criteria: (1) texture interpretations (what is meant by texture); (2) links between surface texture and senses (visual texture vs tactile texture); (3) scale as a determinant of texture (micro-texture vs macro-texture), and (4) the origin of texture (inherent texture vs texturization). The results of the characterization are presented in the first half of the chapter, one criterion at a time, leading to the creation of a surface texture roadmap for product design. The second research objective was to build a typology of texture linked to product functionality. To do this, a photographic analysis of products was made, sourcing product images by visiting retail stores and making online searches. The typology is presented in the second half of the chapter.

5.2 Texture interpretations

Texture is a wide-ranging term used in many different contexts. A vista across a textured landscape is full of content and contrast. When plants are left free to grow over the outside walls of a building, the result can be a texture of natural and synthetic layers. In biting into a gourmet chocolate, the texture provided by each layer of the chocolate, created through contrasting ingredients, provides a rich taste and oral tactual sensation (Fig. 5.1). A textural sound is one that is sonically stationary as opposed to dynamic or evolving (Grill et al., 2011). Within the performing and fine arts, texture is discussed as an element of photographs, paintings, cinematography, and musical performances (Shen et al., 2006). Paintings can have a visual texture with regard

FIGURE 5.1
Biting through the texture of a gourmet chocolate provides multiple tactual sensations, contributing to the enjoyment of eating. © *Owain Pedgley.*

to the layered use of color or shade, but also tactile texture originating from the physicality of the paint itself or the grain of the paper or canvas onto which the paint is applied (Lauer & Pentak, 2015).

In the world of two-dimensional design, texture sits alongside line, shape, color, and value as a basic element (Tersiisky, 2004). The visual weight of a flat surface having a complex texture is said to be greater than that with a simple texture or an absence of texture (Arntson, 1998). In three-dimensional design, texture has increased complexity. Referring to interior design, Mitchell (2019) states that "...textures remind us of nature. They provide the eye with something interesting to look at, but they're also incredibly soothing. (...) Without enough textured elements, a space can feel cold and sterile." Texture can be added to a room by manipulating color, wall coverings, flooring, and furnishings (Martin, 2005; Bowers, 2004). For textiles, the Bauhaus-educated designer Anni Albers wrote of texture as the dominant attribute of woven materials (Albers, 1938). Texture is an integral facet of the physical hand of a textile or fabric, having a great influence on perceived quality. In the world of manufactured products—on which this chapter focuses—textures are associated with the way a product looks on the outside and feels on the surface: commonly referred to as the skin of the product (Menzi, 2010; Lupton, 2002). The personality of a product relies heavily on visual and tactile cues from its surfaces—for which order, proportion, shape, color, and texture are paramount (Ashby & Johnson, 2014). Accordingly, for product designers, surface texture falls within the scope of CMF (colors, materials, and finishes) decision-making and can be used to influence people's product experiences.

Common to all these examples is the unifying principle that texture brings character and interest and therefore has potential to influence all four components of a materials experience (esthetics, meanings, emotions, behaviors). Accordingly, the psychology and cognitive science communities have for decades worked toward understanding people's sensory discrimination of tactile qualities of things, including texture (Tiest, 2010), as well as the vocabulary used to differentiate sensorial qualities of textures, which is often tied directly to surface and bulk material properties (Sakamoto & Watanabe, 2017). For example, through an iterative clustering process, Bhushan et al. (1997) arrived at a set of 98 words to describe different textures (e.g., bumpy, furrowed, matted, scaly, and studded), demonstrating the breadth of vocabulary required to convey different texture qualities.

5.3 Visual versus tactile texture

Vision dominates human information processing and cognition (Hutmacher, 2019), so it is no surprise that the visual properties of product design are prominent in designers' decision-making, leaving tactual properties—and other sensory information—relatively underattended (Zuo et al., 2014; Wastiels et al., 2013; Miodownik, 2005). This said, a huge body of research exists on tactual perception of product surfaces, covering fundamental understanding (mostly cognitive science and psychology research, e.g., Hutmacher, 2019; Sakamoto & Watanabe, 2017; Klatzky & Lederman, 2010; Bhushan et al., 1997; Hollins et al., 1993) as well as specific applications (mostly engineering and design research, e.g., Georgiev et al., 2016; Post et al., 2015; Elkharraz et al., 2014; Jakesch et al., 2011; Chen et al., 2009). Links between surfaces, material properties, and affective response (in users) are common among these studies. Zuo et al. (2014) made a particularly comprehensive study on surface tactility in relation to the user–product interaction, creating a conceptual framework that fuses a material dimension and an experiential dimension, typical of work in the area. The material dimension encompasses objective physical or chemical properties (of the material/product itself), as well as effects of shaping processes and product assembly. The experiential dimension is more subjective, encompassing senses, perception, and user experience, especially with regard to visual, tactile, kinesthetic, and audible effects of material interaction.

From a designer's perspective, texture has been defined as: "tiny marks or shapes in a rather even distribution covering the surface of a shape. These can be slightly irregular or strictly regular, forming a pattern" (Wong, 1993). Note that this definition of texture—as with many others put forward by designers or scholars close to design practice—purposefully avoids constraining texture to a two-dimensional or three-dimensional attribute (if "shape" is interpreted to also include "form"). Nor does the definition imply any hierarchy on visual or tactile experiences from texture. However, distinctions between visual texture and tactile texture are fundamental within the scientific literature, elaborating not only on the physical dimension of texture but also on how texture is practically experienced. These distinctions are important to grasp as a part of foundational discourse on surface textures.

5.3.1 Visual texture

Visual texture is the mainstay of computer visualization of materials (Shen et al., 2006), for example, in computer games, CGI (computer-generated imagery) animations, or CAD (computer-aided design) product development. What is seen on screen is a visually simulated texture, albeit highly convincing. Visual texture on a three-dimensional product gives the impression that there is a tactual effect to be sensed on a surface, but in reality, beyond any inherent texture of the substrate on which the visual texture is applied, no such three-dimensional effect is present. This kind of texture is essentially a graphic finish, also referred to as optical texture or illusory texture. On the surface of a physical product, visual texture has a physically flat presentation. Oftentimes designers use visual textures as analogies to real materials, in which case the visual texture presents as a faux version of the real material. For example, a kitchen countertop can possess the visual qualities of wood grain on an otherwise homogenous and flat thermosetting plastic surface (see Fig. 5.2 for further examples). As demonstrated by the examples, a common way to create visual texture is to reproduce the visual surface impression of real three-dimensional objects or materials. Photographic or digital creation techniques are both applicable. The use of dark and light shading can suggest the shadows and highlights that fall on three-dimensional surfaces. Under certain viewing conditions, visual textures can appear highly realistic and convincing.

5.3.2 Tactile texture

Tactile texture has a volumetric, spatial dimension (Shen et al., 2006). In the world of three-dimensional design, texture enters conversations alongside other prominent design elements such as form, color, material, and weight. The surface of a tactile texture is palpable through the deviations from a smooth uninterrupted surface: the physical peaks and troughs present on the surface can be felt through the skin, most commonly through the fingertips or hands (Lauer & Pentak, 2015). Tactile texture, also referred to as real texture or true texture, is the antithesis of uninterrupted smooth, shiny surfaces (Fig. 5.3). Under strong tangential lighting conditions, tactile texture can cast visually arresting shadows, which in turn affect the perception and personality of the textured surface and product. This raises a crucial point: tactile texture encompasses the provision of both visual and tactual sensorial information (Lauer & Pentak, 2015). Variations

FIGURE 5.2
Visual texture applied to product surfaces, clockwise from top left: sandstone wallpaper, charmed faceted-effect vinyl skin for Xbox-One, leopardskin effect hydrographics, and digitally printed aluminum siding—cherry wood chocolat. *Sandstone wallpaper: © MuralsWallpaper; charmed faceted-effect vinyl skin for Xbox-One: © DecalGirl, Art by FP; leopardskin effect hydrographics: © Liquid Concepts LLC; digitally printed aluminum siding—cherry wood chocolate: © Dizal.*

FIGURE 5.3
Verner Panton classic chair: stunningly curvaceous and absence of surface texture. *Image Creator: Marc Eggimann, © Vitra.*

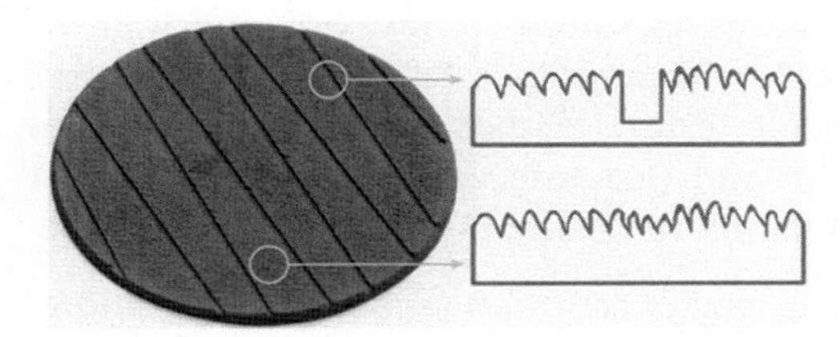

FIGURE 5.4
Macro- and micro-textures on IKEA Pannå coaster. © *Bahar Şener.*

in surface geometry can be assessed not only by the eye—usually as the initial mode of sensory experience—but also by the touch.

5.4 Micro- versus macro-texture

Scale is a crucial matter in texture characterization, as demonstrated through the IKEA Pannå coaster in Fig. 5.4. This simple product is helpful for highlighting an important distinction between what may be termed micro-texture (small scale, appearing as a surface finish/roughness) and macro-texture (large scale, appearing as surface form features). The grooves on the coaster provide a tactile macro-texture, whereas the uneven surrounding surfaces provide a tactile micro-texture. Distinctions between macro- and micro-textures are not so clear in the literature, but parallel distinctions are made between, for example, macro-geometric versus micro-geometric stimuli (Spence & Gallace, 2011); macrostructure versus microstructure (Klatzky & Peck, 2012); coarse versus fine texture (Goldstein & Brockmole, 2016); and large surface features versus microtextural features (Hollins et al., 1993). Product design researchers investigating tactile texture have focused almost exclusively on the micro-scale (Zuo et al., 2016), especially because of its agency to simultaneously improve grip (increase friction) and to disperse light (creating satin and matte visual effects as a contrast between adjacent material surfaces).

A complicating matter is that there is no consensus on the point at which a micro-texture transitions to a macro-texture. For example, as extremes, Jee & Sachs (2000) define macro-texture as submillimetric (<1000 μm wide) surface features, whereas Pardo-Vicente et al. (2019) describe texture in general as an attribute at the level of product form. Klatzky & Lederman (2010) state that texture is a surface geometry property that is distinguishable from the geometry of the host product. They suggest micro-texture transitions to macro-texture if the distance between texture elements (termed as horizontal transition or spatial period) is >200 μm (0.2 mm), but this is not a universally adopted definition. The spatial period is important experientially for tactile textures since it defines how much lateral fingertip or hand movement is needed to perceive a tactual change. Further complicating the perception of tactile texture is uncertainty surrounding the role of protrusion height or depth—and therefore vertical fingertip or hand movement—in sensory discrimination of texture (Tiest & Kappers, 2007).

As a general observation among research studies, tactile textures described at a micrometer (μm) scale usually present as a surface roughness and, therefore, fit both quantitatively and qualitatively into the category of micro-texture. Tactile micro-texture exists as sub- and supersurface disturbances that individually are not easily discernible to the eye, but create an overall surface-wide visual effect (e.g., lustered and matted) and tactual effect (e.g., roughened). Practically, micro-texture is realized at the level of material surface finishes, through industrial processes that either roughen or deposit a coating, for example, mold tool spark erosion, shot-blasting, splattering, and powder coating (Brainard, 2005). The tactility of a micro-texture is conventionally evaluated on a smooth-rough perceptual scale; in essence, a scale that tends toward macro-texture as roughness perception increases (Hollins et al., 1993).

In contrast, tactile macro-texture as a term seems best reserved for relatively small-scale surface form features that are obvious to the eye (>1 mm wide, but often much wider depending on the area of the surface to be textured), with at least as wide—often much wider—spatial period. See, for example, the textured surfaces of products showcased in the Form Fächer guide (Zurich University of The Arts, 2009), which exemplify this definition. Macro-texture form features may be variously described as bulges, ridges, ribs, protrusions, and raised edges (examples of supersurface, embossed, elevated, protruding, or high relief features) or dents, recesses, furrows, channels, and dimples (examples of subsurface, debossed, depressed, sunken, or low relief features). Blind holes, depending on their abundance on a surface, may create a form of low relief texture. However, thru-holes, perforations, and grilles, where a surface is pierced (Fig. 5.5), are not strictly within the scope of texture, even though running one's finger across the surface would reveal a series of troughs disrupting an otherwise flat surface. The tactility of macro-textures has been evaluated on a flat-bumpy perceptual scale (Hollins et al., 1993).

5.5 Inherent texture versus texturization

It is often the case that a material possesses micro-texture or macro-texture that is attributable either to underlying material properties or to ways in which a material is processed into a semifinished material (for subsequent use in a product) or directly into an artifact. In such cases, a texture is said to be inherent to the material or its processing, for example, the relief and cracks associated with leather, the

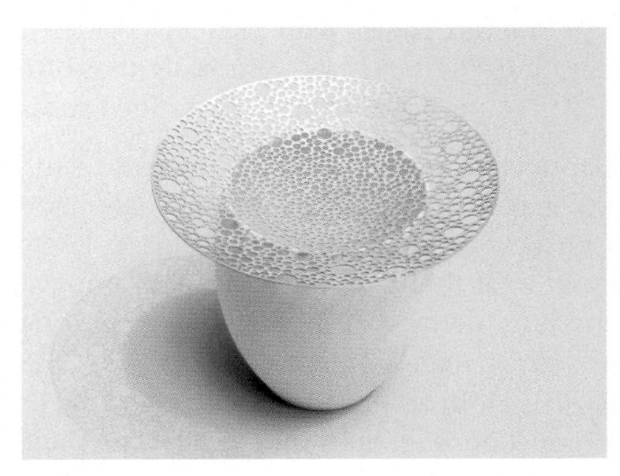

FIGURE 5.5
Queen Anne's Lace tea strainer from etched silver, Ted Muehling Studio: beautiful, but through holes are not strictly a type of texture. © *Loring McAlpin.*

FIGURE 5.6
Inherent material textures, clockwise from top left: gray flock on tan fabric, leather bag and belt, Stone Forest Wabi Vessel washbasin, mycelium and timber stool (developed by designer and maker, Sebastian Cox and design thinker, Ninela Ivanova), synthetic fleece, and fallen palm leaf round bowl. *Gray flock on tan fabric: © New Creation Inc.; leather bag and belt: © Dean Neitman | Dreamstime.com; Stone Forest Wabi Vessel washbasin: © Stone Forest; mycelium and timber stool: Photograph by Petr Krejci; synthetic fleece: © Berghaus; fallen palm leaf round bowl: © Bionatic/Naturally Chic.*

hand of piled fabric, or the hard undulations of carved stone (Fig. 5.6). The processing of concrete in architecture, for example, can lead to surface textures that mirror the formwork shuttering used, such as wooden planks with pronounced wood grain, or smooth panels with regularly placed ties (Hegger et al., 2007). Grown materials and DIY-Materials—especially those formed of bio-ingredients (Rognoli et al., 2015)—can have remarkably textured surfaces.

Familiarity with textures that are inherent to materials and their processing is an important part of the designer's materials knowledge and may create a launch point for new product designs (Alesina & Lupton, 2010). Furthermore, material aging and wear through product use can bring about changes to inherent textures or introduce new textures that were not originally present, such as peeling finishes, crease marks, or dents (Lilley et al., 2019; Pedgley et al., 2018; Rognoli & Karana, 2014). In contrast to inherent texture, texturization refers to the purposeful introduction of texture through a design and application process. In essence, texturization results in a texture on a material surface that would not be present without a concerted effort by the designer to plan and realize such a texture.

5.6 Surface texture roadmap

Fig. 5.7 provides a roadmap that summarizes the surface texture characteristics reviewed in phase 1 of the *textureface* project. The roadmap provides a new practical starting point for discussion and action on product surface textures among designers and design researchers.

5.7 Functional texture

The second research objective for phase 1 of the *textureface* project was to build a typology of texture tied to product functionality. The work commenced with a field study visiting retail stores in the United Kingdom and Turkey to photograph surface textures on products, supplemented by sourcing of online images. The photographs and images combined to form a diverse database of textured product examples falling within the roadmap category {purpose = functional texture; senses = tactile texture; scale = macro-texture; origin = texturization}. Diversity in product examples ($n = 200+$) was achieved through the choice of stores (e.g., department stores selling products across many sectors) and careful choice of search terms for Google's image search engine. The photographs and images were compiled into montages and cut-and-paste posters to aid multiple rounds of cross-comparison and analysis by the research team. Two criteria were used to construct the typology:

- Functionality. How does the texture contribute to—or deliver—product function?
- Topography. What parameters describe the variation of geometries/forms for functional texture?

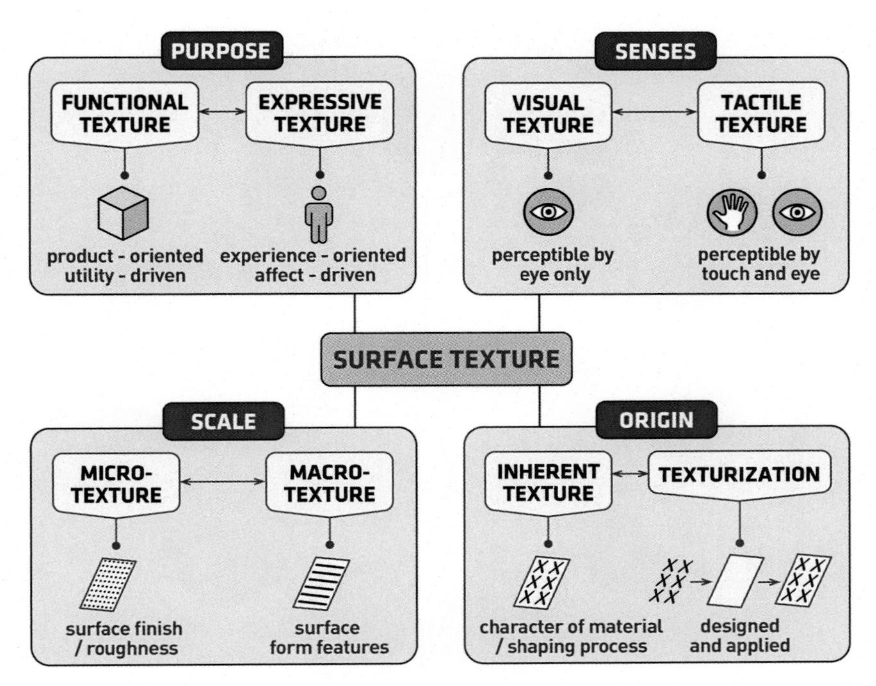

FIGURE 5.7
Surface texture roadmap. © *Bahar Şener/Owain Pedgley.*

5.7.1 Analysis by functionality

In total, 11 categories of functional texture were identified, creating a typology of functional texture visualized in Fig. 5.8. In some cases, the texture achieves the main function for the product; in other cases, it provides a supporting role. To assist the categorization, the user–product and product–product domains of interaction by Sener & Pedgley (2019) were used.

Under the user–product interaction category, people purposefully try to achieve something through interacting with a product (instrumental interaction) and the texture assists the achievement of the task. The subcategory Fingertip Grip describes cases where texture supports dexterous and precision (pinch) grip, allowing users to make a controlled movement of a component (e.g., twist, slide, and push) by providing a secure purchase on the surface of that component with the fingertip. Examples include rotary controls, rotary demountable components (e.g., bottle tops) and sliding switches, controls, or catches (e.g., craft knife). Oppositely, a secure purchase may also be required where movement should be prevented—in other words, in cases where fingertip interaction should result in a firm grip. Examples include pen and pencil shafts, hand tools, and handheld products where grip is imperative for security or safety (e.g., razors).

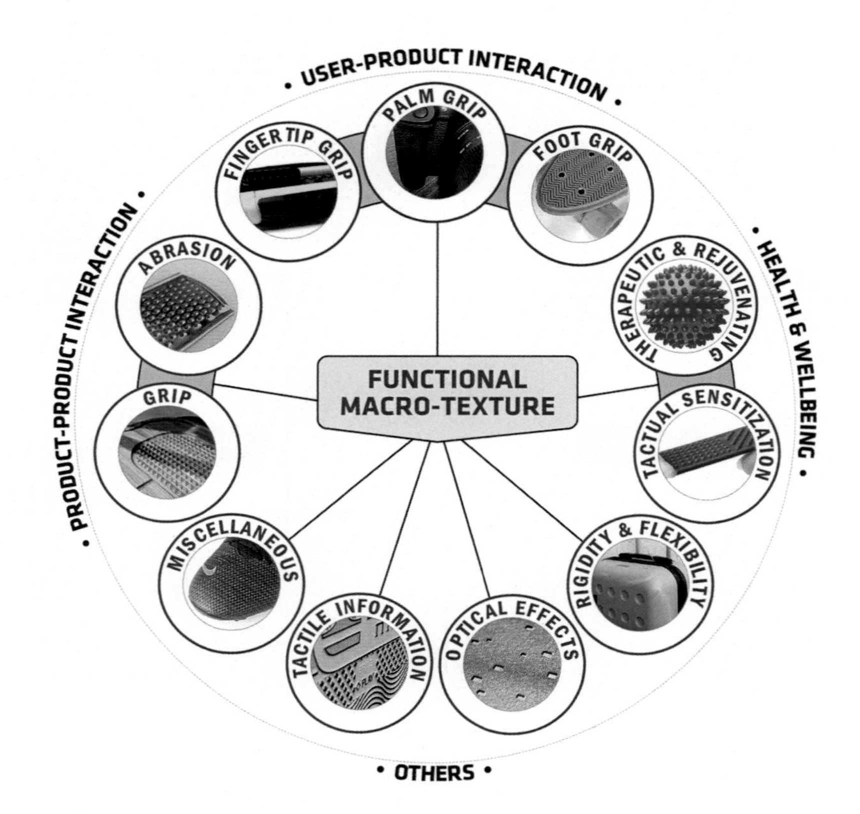

FIGURE 5.8

Typology of functional macro-texture, with examples: grip (chopping board), abrasion (cheese grater), fingertip grip (craft knife), palm grip (camera and lens), foot grip (penny board), therapeutic & rejuvenating (massage ball), tactual sensitization (flat textured spoon), rigidity & flexibility (suitcase), optical effects (flat glass), tactile information (tactile map), and miscellaneous (football boot). *© Bahar Şener/Owain Pedgley. Use of flat textured spoon image granted by Ark Therapeutic; use of tactile map image granted by ClickAndGo Wayfinding.*

The subcategory Palm Grip describes cases where texture contributes to the achievement of a grasping, grabbing, or holding power grip, during which a product can be held securely without slippage or movement. In this subcategory, there is a blurring between the contribution of texture toward achieving grip and the contribution toward visual esthetics on relatively large surfaces. For example, when texture is located on a handle, on a control, or on a control/grip surface, the contribution is more obviously functional. When texture is located on the main body, away from areas of interaction, the contribution to function is less obvious. In the latter case, the texture can offer improved grip to pick up and hold a product securely, but this is a rather tenuous connection since it is carried out rarely and not a part of instrumental interaction. Instead, for the majority of the time, such texture is appreciated as something to behold and appreciate visually. Examples include artifacts that may occasionally be moved (e.g., vases, pots, and jars); containers with more frequent interaction (e.g., drinks glasses, drinks bottles, perfume bottles, teapots, and mugs); textured covers over smooth product surfaces (e.g., mobile phone cases); control/grip surfaces on products (e.g., shavers and kitchen gloves); handles (e.g., racquets, baby strollers, bicycles, handheld products, steering wheels, torches, and toothbrushes); fragile items needing to be transported or carried securely (e.g., cameras and portable hard disks); and oversized controls on products (e.g., chair tilt adjustment and water bottle lids).

The subcategory Foot Grip describes cases where texture contributes to achieving a secure connection between somebody's foot and the surface of a product, for the purpose of preventing movement (nonslip). Examples include bathmats; flooring; product foot controls (e.g., car pedals and pedal bin); and standing surfaces (e.g., skateboard).

Under the Product–Product Interaction category, people purposely try to achieve something through the action of one product on another product or object. In other words, functionality is achieved through surface texture interactions. The surfaces involved do not have direct user intervention. The subcategory Grip describes cases where the functionality is related to maintaining control, grip, and ensuring no slippage between surfaces. Examples include clothing (on shoulders, to prevent rubbing and movement of accessories such as rucksacks); storage areas, shelving, and coasters (to prevent rolling or movement of placed items); chopping boards (to keep food still during cutting); jar openers (combined with high friction materials to provide extra secure grip); toys (to provide friction fit to join parts, such as Lego); and shoe outer soles and tire treads (to provide grip and lift water).

The subcategory Abrasion covers a special case where texture is realized in a durable material for the specific purpose of providing abrasive (material removal) properties. Examples include sheet metal textures (in cheese/food graters); ground metal textures (in workshop files); and ground glass textures (in nail files).

Under the Health & Well-being category, texture is provided on the surface of products for the purpose of stimulating the skin and muscles. The subcategory Therapeutic & Rejuvenating describes cases related to relief and relaxation, for example, from the application of pressure onto muscles/skin (e.g., foam exercise rollers, massage balls, baby and pet teethers) and through to gentle touch (e.g., massaging face mittens to promote skin renewal).

In the subcategory Tactual Sensitization, texture serves to stimulate or regulate the tactile sense. For example, feeding spoons that are textured to provide added tongue stability and awareness at eating time for babies transitioning from puréed to solid/textured foods; sensory play toys for babies, having variety in textures and material properties to simulate brain development and develop the sense of touch; and sensory sticks, textured to treat sensory (tactual) processing disorders such as hyperesthesia (oversensitivity, leading to tactile defensiveness) or hypoesthesia (undersensitivity, leading to tactile seeking).

Under the Others category, texture is put to a wide variety of use. The subcategory Rigidity & Flexibility covers the application of textures to increase or decrease the structural stiffness of a part (increasing or decreasing the second moment of area). Examples of increased rigidity through texturization—which often involves much larger than millimetric features—include metal panels on cars or white goods, plastic panels on suitcases, silicone cake molds, and polyethylene terephthalate (PET) water bottles that otherwise have poor structural properties. Examples of increased flexibility through texturization include creating localized subsurface textures on gloves and indentations on watch straps.

As raised earlier in the chapter, texture on a micro-scale is often used to create optical effects such as visual contrasts and matte finishes as an alternative to high gloss. The subcategory Optical Effects covers the use of macro-texturization to achieve functional optical effects. Examples include obscured flat glass panels for privacy (such as used for internal walls, shower cubicles, and door panels); texturized glass or transparent plastic in car headlight fixtures, to help distribute light from the point source of a bulb; and texturized plastic car dashboards, giving low levels of reflection and reduced glare to avoid driver distraction.

The subcategory Tactile Information describes the use of texture as an alternative to graphical/verbal instruction. Examples include braille—surface texture perceived by the fingertips that is organized into a language alternative to words; and other applications to assist the visually impaired, such as bumps on pavement tiles to help wayfinding, with different texture designs and directions indicating different information.

The final subcategory Miscellaneous captures functional uses of texture that do not fit into the preceding categories. Products in this category include paper coffee cups (relief on the cup surface traps air for insulation and reduces surface contact area to avoid burning hands); silicone cake molds (dots on the base surface help release cakes easily); paper towels (texture increases paper volume and improves absorbency); cotton blankets (embossing gives softness and structural stability); roller paint trays (ridges control movement of the liquid paint); golf balls (dimples provide aerodynamic control); golf clubs (grooves provide aerodynamic control); football boot uppers (texture gives traction for ball contact and aerodynamic control); and heat sinks (texture increases surface area to dissipate heat from components such as LED lights and audio amplifiers).

5.7.2 Analysis by topography

Analysis of the 200 + textured products revealed some common descriptive points that can be used to help describe surface texturization in a systematic manner. The descriptions are presented under three categories: Geometry, Bipolar Scaling, and Arrangement. Note that these descriptions are at a general level and are not specific to any product sector. Effects of topography on functionality are not implied, but future research studies could use the descriptions as texturization variables to measure effects on, for example, usability, task achievement, or visual and tactile attractiveness within specific product sectors.

Under Geometry, macro-texture form elements can be defined as follows.

- Extruded polygon, for example, circle (dot), triangle, square, rectangle, diamond, and hexagon;
- 3D volume, for example, prism, pyramid, hill, dome, and scallop (inverse dome);
- Longitudinal, for example, notch, ridge, channel, strip, and lozenge;
- Irregular, for example, globular, cellular, emblem, figure (heart, star, etc.), wave, material imitation, and decorative pattern.

Obvious differences between macro-textured surfaces can be defined using Bipolar Scaling (semantic differential scaling) as follows:

- Single size elements—multisize elements
- Small relative size—large relative size
- High-density elements—low-density elements
- Subsurface—supersurface (exception: faceted textures create a fully undulating surface)
- Flat surface application—curved surface application
- Sharp edge elements - smooth (filleted) edge elements

Arrangement refers to other dimensions that influence the final macro-texture.

- Boundary. Defines the extent/edge of texturization on the product surface (sometimes a texture is gradually tapered or faded out to flat, rather than ending abruptly).
- Tessellation. Defines the way in which the texture elements are brought together, often using geometrical arrangements (e.g., X-Y spacing, linear placement, radial placement, and concentric placement) or with irregularity.

5.8 Discussion and conclusions

Successful product design is frequently attributed to excellent functionality combined with satisfaction and pleasure from appearance and interaction

(Norman, 2013). As the topics of discussion have shown throughout this chapter, material surface texture has the potential to influence product success against all these criteria. The chapter has worked toward the use of purposefully designed and applied textures to supplement or extend product functionality in manufactured products. In such cases, surface texturization often provokes actions from users, directly addressing the performative aspects of materials experience.

Such functional gains cannot be considered in isolation; as is inevitable in design, those functional enhancements have implications for sensorial-expressive effect. Wider esthetic, emotional, and meaningful experiences must still be considered as part of the functional texture concept. In this regard, a subset of the product examples listed in the chapter can be regarded as possessing "functional pleasant tactility" (Fennis, 2012), which describes a situation where designers have made an effort for a surface texture to feel tactually pleasant in use as a cooperative attribute to the functional gains brought by the texture. Texture can also be used playfully. Rognoli (2015) gives an example of how designers are adept at working with texture at different levels: for example, creating macro-texturization through embossing the leather on the outside of a handbag, which in turn mirrors the smaller scale texture of the stitching on the inside of the bag. Texturization can still, of course, be conceived without any particular functionality in mind. From this perspective, the texture characterization criteria presented earlier in the chapter, and summarized in the surface texture roadmap, are informative and relevant to designing what may be termed expressive or decorative textures.

As observed from the analysis of 200 + products, texture is often used to enhance the grip of products. To amplify the nonslip effects of texture, designers sometimes specify materials having a high coefficient of friction, such as thermoplastic elastomers or rubbers. Stylistically, this often results in high-grip components that are visually strongly discernable from the main body of a product, both regarding color choice and texture. Coinjection molded toothbrushes are a very well-known example of this phenomenon. In more demanding contexts, such as the nonslip grips of sports products, surface texturization can be a fruitful area of research bridging sporting performance, design research, and macro-scale tribology.

The physicality of tactile texture is important from the point of view of providing product durability. When considering the Tactile Information subcategory from the typology of functional textures, then unlike applied graphics, texture is not at risk of rubbing off over time. Furthermore, the physicality of texture gives rise to the possibility of blind interaction: in situations where gazing at the point of interaction is not possible or desirable, functional textures can guide us tactually about where to grip or interact with a product.

One example can be in the use of textured driver controls out of the line of vision, building upon principles laid out for highly tactile interfaces reliant predominantly on form (Porter et al., 2005). Continuing with the automotive theme, trends in automotive human–machine interaction/interface (HMI) have seen a gradual increase in the size and reliance of touchscreens to access functionality, away from dedicated tangible controls. This is a consequence of the general processes of dematerialization, involving a shift from physical products to digital apps (van Campenhout et al., 2013). Nevertheless, a balance in interaction modalities is often aimed for in highly interactive environments such as automotive HMI—mixing the digital and the physical—in which case, textures on tangible controls or interior surfaces can provide a "design antidote" (Pond, 2014) to the materially diminished touch experiences of touchscreens and high gloss surfaces. Playful interactions can be achieved through textures: for example, several concept designs exist for scratchable controllers, where scratching a texture creates a sound signature which in turn is mapped to activating/deactivating certain product functions (Murray-Smith et al., 2008).

It is worth mentioning at this point the related emerging field of tactile displays, which can be regarded as a dynamic extension of surface macro-texturization. Tactile displays possess changeable surface forms or textures, which on a macro-scale can be achieved through smart materials, actuators, and computational materials. The result is a surface that can change its tactile character, either automatically or by intervention. Jung et al. (2010) made inflatable cup and tactile mouse conceptual designs based around these principles. Metamaterials defined as materials possessing a designed and geometrically complex internal microstructure that can controllably deform and reshape through various means (e.g., motors, pneumatics, shape memory polymers, electromagnets, thermal response, light-activation, and hydrophilic action) have also been demonstrated as a route to creating dynamic surface textures, which can influence action tendency (Ion & Baudisch, 2020; Ion et al., 2018). In another emerging field, texture is explored through the medium of biomaterials, for example, by encouraging plant roots to grow into intricate repeated structures (see, e.g., grown products by Diana Scherer showcased in Karana, 2020). Such applications can be conceived with functional or expressive applications in mind.

Surface texture has been known for some time to be highly suited to direct materialization via 3D-printing methods. As such, it offers a plausible route, among others, to realize mass customization (Campbell et al., 2003). In recent years there has been a large expansion in practical advice about how to model complex textures using industry-standard software (e.g., SolidWorks, Rhino) and visual programming plug-ins (e.g., Grasshopper) paying attention to the 3D printability of the texture. As the resolution of

3D printers becomes greater, so too does the ability to create one-off surface textured artifacts or master patterns for volume production, including at the macro–micro-scale boundary.

For all of the benefits that surface textures can bring, there are still negative effects that must be striven to be avoided. Textured surfaces are prone to catching dirt and can therefore be harder to clean and maintain good hygiene compared with nontextured surfaces. Durability can also be a concern, if the materialized texture is fragile or susceptible to getting knocked.

5.9 Conclusion

Materials are often processed or formed to leave a surface texture that adds value to a product. If texture does not come inherently, through the material itself or through its shaping and finishing, then the designer must consciously texturize the material surface. This chapter has reported on the first phase of the three-phase *textureface* project, making a detailed examination of the characteristics of surface texture and a review of how texturization of an otherwise plain surface can be functionally beneficial. The presented surface texture roadmap provides a launch point for designers and design researchers to consider how to use surface texture in product designs.

The *textureface* project has also provided the first systematic analysis of how texturization can deliver or support product function. Eleven categories of functional texture were identified as a typology, providing informative and inspirational examples for designers to use texturization to support or deliver product function. The chapter has contributed an investigation of the pragmatic uses of texture, which complements the far more commonly discussed sensorial-expressive qualities of materials and texture, mostly concerned with user affect and designer expression.

Acknowledgments

Funding for the work reported in this chapter was received from the BAP Project (Scientific Research Projects) scheme of Middle East Technical University, under Grant No. BAP-02-03-2014-001 "Research into 3D Surface-Texturization for Functional and Hedonic Product Enhancement." Thanks are extended to Yavuz Paksoy for his involvement. A follow-up work package conducted at the University of Liverpool, School of Engineering, entitled "Exploration and Application of 3D Surface-Texturization for Functional and Hedonic Product Enhancements" was funded under the Brazilian student mobility scheme *Science Without Borders*. Thanks are extended to Karolyne Santos de Castro for her involvement. Thanks are also expressed to students of ENGG220 Product Development 2 course at the University of Liverpool, who designed textured travel mugs and textured chocolates under the authors' guidance, helping to test out many of the principles put forward in the chapter.

References

Albers, A. *Work with material.* (1938). <https://albersfoundation.org/artists/selected-writings/anni-albers/> Accessed 17.06.20.

Alesina, I., & Lupton, E. (2010). *Exploring materials: Creative design for everyday objects.* New York: Princeton Architectural Press.

Arntson, A. (1998). *Graphic design basics.* Fort Worth: Harcourt Brace College Publishers.

Ashby, M., & Johnson, K. (2014). *Materials and design* (3rd ed.). Oxford: Butterworth-Heinemann.

Bakker, S., de Waart, S., & van den Hoven, E. (2015). Tactility trialing: Exploring materials to inform tactile experience design. In *Proceedings of DeSForM 2015* (pp. 119–128). Milan: Politecnico di Milano.

Bhushan, B., & Sayer, R. (2007). Surface characterization and friction of a bio-inspired reversible adhesive tape. *Microsystem Technology, 13*(71), 71–78.

Bhushan, N., Rao, A. R., & Lohse, G. L. (1997). The texture lexicon: Understanding the categorization of visual texture terms and their relationship to texture images. *Cognitive Science, 21*(2), 219–246. Available from https://doi.org/10.1207/s15516709cog2102_4.

Bowers, H. (2004). *Interior materials and surfaces: The complete guide.* Ellicott Station: Firefly Books.

Brainard, S. (2005). *A design manual* (4th ed.). London: Pearson.

Camere, S., Schifferstein, H. N. J., & Bordegoni, M. (2018). From abstract to tangible: Supporting the materialization of experiential visions with the experience map. *International Journal of Design, 12*(2), 51–73.

Campbell, R., Hague, R., Sener, B., & Wormald, P. (2003). The potential for the bespoke industrial designer. *The Design Journal, 6*(3), 24–34.

Chen, X., Barnes, C., Childs, T., Henson, B., & Shao, F. (2009). Materials' tactile testing and characterization for consumer products' affective packaging design. *Materials and Design, 30*(10), 4299–4310.

Elkharraz, G., Thumfart, S., Akay, D., Eitzinger, C., & Henson, B. (2014). Making tactile textures with predefined affective properties. *IEEE Transactions on Affective Computing, 5*(1), 57–70. Available from https://doi.org/10.1109/T-AFFC.2013.21.

Fenko, A., Schifferstein, H. N. J., & Hekkert, P. (2010). Shifts in sensory dominance between various stages of user-product interactions. *Applied Ergonomics, 41*(1), 34–40.

Fennis, T. (2012). *Exploring and implementing pleasant touch in the interface of products for design purposes: The case of a Bang & Olufsen remote control* [Unpublished MSc thesis, Department of Industrial Design, Middle East Technical University/Faculty of Industrial Design Engineering, Delft University of Technology].

Fratzl, P. (2007). Biomimetic materials research: What can we really learn from nature's structural materials? *Journal of The Royal Society Interface, 4*(15), 637–642. Available from https://doi.org/10.1098/rsif.2007.0218.

Georgiev, G., Nagai, Y., & Taura, T. (2016). Modelling tactual experience with product materials. *International Journal of Computer Aided Engineering and Technology, 8*(1–2), 144–163.

Goldstein, E., & Brockmole, J. (2016). *Sensation and perception* (10th ed.). Belmont: Wadsworth.

Grill, T., Flexer, A., & Cunningham, S. (2011). Identification of perceptual qualities in textural sounds using the repertory grid method. In *Proceedings of the 6th audio mostly conference: a conference on interaction with sound* (pp. 67–74). Coimbra: ACM.

Hassenzahl, M. (2003). The thing and I: Understanding the relationship between user and product. In M. Blythe, K. Overbeeke, A. Monk, & P. Wright (Eds.), *Funology: From usability to enjoyment* (pp. 31–42). Dordrecht: Kluwer Academic Publishers.

Hegger, M., Drexler, H., & Zeumer, M. (2007). *Basics materials*. Basel: Birkhäuser.

Herssens, J., & Heylighen, A. (2010). Haptic design research: A blind sense of place. In *Proceedings of ARCC/EAAE international conference on architectural research* (n.p.). Washington, DC: Architectural Research Centers Consortium.

Hollins, M., Faldowski, R., Rao, S., & Young, F. (1993). Perceptual dimensions of tactile surface texture: A multidimensional scaling analysis. *Perception and Psychophysics, 54*(6), 697–705.

Hutmacher, F. (2019). Why is there so much more research on vision than on any other sensory modality? *Frontiers in Psychology, 10*(2246), 1–12. Available from https://doi.org/10.3389/fpsyg.2019.02246.

Ion, A., & Baudisch, P. (2020). Interactive metamaterials. *Interactions*, 88–91. Available from https://doi.org/10.1145/3374498.

Ion, A., Kovacs, R., Schneider, O., Lopes, P., & Baudisch, P. (2018). Metamaterial textures. In *Proceedings of the 2018 CHI conference on human factors in computing systems* (pp. 1–12), No. 336. doi:10.1145/3173574.3173910.

Jakesch, M., Zachhuber, M., Leder, H., Spingler, M., & Carbon, C.-C. (2011). Scenario-based touching: On the influence of top-down processes on tactile and visual appreciation. *Research in Engineering Design, 22*, 143–152.

Jee, H., & Sachs, E. (2000). Surface macro-texture design for rapid prototyping. *Rapid Prototyping Journal, 6*(1), 50–60. Available from https://doi.org/10.1108/13552540010309877.

Jung, H., Altieri, Y., & Bardzell, J. (2010). SKIN: Designing aesthetic interactive surfaces. In *Proceedings of fourth international conference on tangible and embedded interaction TEI2010* (pp. 85–92), Cambridge: ACM.

Karana, E. (2020). *Still alive: Livingness as a material quality in design*. Breda: Avans University of Applied Sciences.

Karana, E., Pedgley, O., & Rognoli, V. (2014). *Materials experience: Fundamentals of materials and design*. Oxford: Butterworth-Heinemann.

Klatzky, R., & Lederman, S. (2010). Multisensory texture perception. In M. J. Naumer, & J. Kaiser (Eds.), *Multisensory object perception in the primate brain* (pp. 211–230). New York: Springer.

Klatzky, R., & Peck, J. (2012). Please touch: Object properties that invite touch. *IEEE Transactions on Haptics, 5*(2), 139–147.

Lauer, D., & Pentak, S. (2015). *Design basics* (9th ed.). Belmont: Wadsworth.

Lilley, D., Bridgens, B., Davies, A., & Holstov, A. (2019). Ageing (dis)gracefully: Enabling designers to understand material change. *Journal of Cleaner Production, 220*, 417–430.

Lupton, E. (2002). *Skin: Surface, substance and design*. London: Laurence King.

Martin, C. (2005). *The surface texture bible*. New York: Harry N. Abrams.

McCardle, J. (2015). Science informed design: Involving the physical and natural sciences. In *Proceedings of international conference on engineering and product design education* (pp. 1–6). Loughborough: Loughborough Design School.

Menzi, R. (Ed.), (2010). *Make up: Designing surfaces*. Ludwigsburg: AV Edition.

Miodownik, M. (2005). A touchy subject. *Materials Today, 8*(6), 6.

Mitchell, N. (2019). Designer secrets: Transform your space with texture. *Apartment Therapy*. <https://www.apartmenttherapy.com/designers-secrets-5-ways-to-add-texture-to-a-room-192202> Accessed 16.06.20.

Murray-Smith, R., Williamson, J., Hughes, S., Quaade, T., & Strachan, S. (2008). Rub the stane. In *Proceedings of CHI'08* (pp. 2355–2360). doi:10.1145/1358628.1358683.

Norman, D. (2013). *The design of everyday things*. Cambridge: MIT Press.

Özcan, N. (2008). The emotional reactions of tactual qualities on handheld product experiences [Unpublished MSc thesis, Department of Industrial Design, Middle East Technical University].

Pardo-Vicente, M., Rodriguez-Parada, L., Mayuet-Ares, P., & Aguayo-Gonzalez, F. (2019). Haptic hybrid prototyping (HHP): An AR application for texture evaluation with semantic content in product design. *Applied Sciences*, *9*(5081), 1–20.

Pedgley, O., Sener, B., Lilley, D., & Bridgens, B. (2018). Embracing material surface imperfections in product design. *International Journal of Design*, *12*(3), 21–33.

Pond, P. *Texture and touch is crucial in product to consumers today*. (2014). <https://www.linkedin.com/pulse/20141102173020-33624313-texture-touch-is-crucial-in-product-to-consumers-today/> Accessed 09.06.20.

Porter, J.M., Summerskill, S., Burnett, G., & Prynne, K. (2005). BIONIC—'eyes-free' design of secondary driving controls. In L. Gibson, P. Gregor, & D. Sloan (Eds.), *Accessible design '05: Proceedings of the 2005 international conference on accessible design in the digital world* (n.p.). Swindon: BCS Learning & Development Ltd.

Post, R., Saakes, D., & Hekkert, P. (2015). A design research methodology using 3D-printed modular designs to study the aesthetic appreciation of form and material. In *Proceedings of DeSForM 2015* (pp. 347–350). Milan: Politecnico di Milano.

Rognoli, V. (2015). Dynamic and imperfect as emerging material experiences: A case study. In *Proceedings of DeSForM 2015* (pp. 66–76). Milan: Politecnico di Milano.

Rognoli, V., Bianchini, M., Maffei, S., & Karana, E. (2015). DIY materials. *Materials and Design*, *86*, 692–702.

Rognoli, V., & Karana, E. (2014). Toward a new materials aesthetic based on imperfection and graceful aging. In E. Karana, O. Pedgley, & V. Rognoli (Eds.), *Materials experience* (pp. 145–154). Oxford: Butterworth-Heinemann.

Sakamoto, M., & Watanabe, J. (2017). Exploring tactile perceptual dimensions using materials associated with sensory vocabulary. *Frontiers in Psychology*, *8*(569), 1–10.

Sener, B., & Pedgley, O. (2019). Accelerating students' capability in design for interaction. In *Proceedings of DRS Learn-X Design 2019* (n.p.). Ankara: Middle East Technical University.

Shen, J., Jin, X., Mao, X., & Feng, J. (2006). Completion-based texture design using deformation. *Visual Computing*, *22*, 936–945.

Sonneveld, M., & Schifferstein, H. (2008). The tactual experience of objects. In H. Schifferstein, & P. Hekkert (Eds.), *Product experience* (pp. 41–67). Oxford: Elsevier.

Spence, C., & Gallace, A. (2011). Multisensory design: Reaching out to touch the consumer. *Psychology and Marketing*, *28*(3), 267–308.

Tersiisky, D. *The elements and principles of design*. (2004). <http://metalab.uniten.edu.my/~ridha/PrinCiplesOf_Design/references/design.pdf> Accessed 16.06.20.

Tiest, W. (2010). Tactual perception of material properties. *Vision Research*, *50*(24), 2775–2782.

Tiest, W., & Kappers, A. (2007). Haptic and visual perception of roughness. *Acta Psychologica*, *124*(2), 177–189.

van Campenhout, L., Frens, J., Overbeeke, K., Standaert, A., & Peremans, H. (2013). Physical interaction in a dematerialized world. *International Journal of Design*, *7*(1), 1–18.

Wastiels, L., Schifferstein, H., Wouters, I., & Heylighen, A. (2013). Touching materials visually: About the dominance of vision in building material assessment. *International Journal of Design*, *7*(2), 31–41.

Wong, W. (1993). *Principles of form and design*. New York: Van Nostrand Reinhold.

Zuo, H. (2010). The selection of materials to match human sensory adaptation and aesthetic expectation in industrial design. *METU Journal of the Faculty of Architecture*, *27*(2), 301–319.

Zuo, H., Hope, T., & Jones, M. (2014). Tactile aesthetics of materials and design. In E. Karana, O. Pedgley, & V. Rognoli (Eds.), *Materials experience* (pp. 27–37). Oxford: Butterworth-Heinemann.

Zuo, H., Jones, M., Hope, T., & Jones, R. (2016). Sensory perception of material texture in consumer products. *The Design Journal, 19*(3), 405–427.

Zurich University of The Arts. (2009). *Form fächer: Form guide*. Ludwigsburg: AV Edition.

CHAPTER 6

Material change: transforming experience

Debra Lilley[a], Ben Bridgens[b]
[a]School of Design and Creative Arts, Loughborough University, Loughborough, United Kingdom, [b]School of Architecture, Planning and Landscape, Newcastle University, Newcastle upon Tyne, United Kingdom

6.1 Introduction

Scenario 1: "You open the door of your rental car; it's sunny, the car is hot and humid, the smell of stale cheese and onion crisps hits your nostrils full force. You feel slightly nauseous. The steering wheel is sticky, residue from the previous user's fizzy drink. You think about reaching for the wet wipes to attempt to remove the thin layer of sweaty film from the hot plastic surface, but why bother? Your time in this vehicle is fleeting—a means to get from A to B. You reach your hotel. It is bland, clinical, a sea of homogeneous white towels and crisp cold bed sheets. You could be anywhere. The pillows are flat but serviceable. You sit down at a small desk, the surface of a thin layer of wood veneer chipped on one corner to reveal the low-cost chipboard within. You drop your car keys onto the desk with a clatter; the already scratched surface and cheap construction does little to elicit your respect. You reach down for your leather satchel and feel a sense of calm as your eyes work their way over the familiar textures and patterns of the worn leather, the deep scratch that vividly reminds you of a close-call with a rhinoceros in Mozambique, the suppleness of the leather gained through years of care and nourishment. You pull out a well-thumbed book, a gift from an anonymous fellow traveler keen to facilitate its onward passage. The spine is creased, scrawled notes emphasize meaningful passages, reflections on words you read afresh with the lens of others before you."

We are constantly interacting with the material world around us, and these interactions, in turn, change the natural and man-made materials embodied in our streetscapes, buildings, and products. Our relationship with materials, and the materials themselves, are constantly changing. Designers and manufacturers typically focus on the pristine object that entices the purchaser, with little consideration of how materials will change over days, years, or centuries of use

Materials Experience 2. DOI: https://doi.org/10.1016/B978-0-12-819244-3.00027-2

(Nobels et al., 2015), and how in turn this will influence the product's life: from careful use to abuse and from reuse to disposal.

Since the Industrial Revolution, mechanized "perfection" has steadily replaced craftsmanship. In *The Nature and Art of Workmanship* David Pye (1968) describes "workmanship of risk" "in which the quality of the result is not predetermined, but depends on the judgement, dexterity and care which the maker exercises as he works" (p. 20). In contemporary society workmanship of risk has been replaced by "workmanship of certainty." A combination of homogeneous man-made materials (such as plastics, metals, and glass) and an array of increasingly sophisticated, automated manufacturing processes have enabled designers to deliver immaculately smooth, indistinguishable objects. But from the moment of purchase the surface of these objects is exposed to a diverse range of stimuli including heat, light, moisture, wear, and impact. A complex interaction of physical, chemical, and biological processes results in changes to the material surface, altering both the look and feel of the object (Fig. 6.1). Environmental stimuli include moisture,

FIGURE 6.1
Materials change. Clockwise from top left: a plastic spade is severely faded by sunlight despite being designed for outdoor use; paint wears away from floor boards, revealing the grain of the wood and highlighting the routes that people walk; denim fades from dark blue to white with repeated washing and use, this appearance is highly valued and many products are "preaged" by the manufacturer by stone washing; white limestone blocks have slowly been colored by water running down from a copper roof above.

light, temperature, growth of mold and fungi, and reaction with oxygen and other chemicals in the atmosphere. Physical interaction includes handling, carrying, and dropping an object resulting in impact, ablation (chipping of the surface), abrasion (scratching and polishing), and accumulated dirt. There is also deliberate interaction with the material surface, which ranges from care (cleaning, nourishing with wax or oil, and polishing) to repair and personalization. These stimuli combine to create a complex surface patina, which varies both spatially (across the surface of an object) and temporally.

Materials communicate to us on many levels signaling their provenance, worth, and capabilities. The patina or "traces of use" (Giaccardi et al., 2014) evident in their surfaces are "a type of material history ... inscribing a unique and personal semantic narrative into the objects through material experiences" (Odom & Pierce, 2009, p. 3796). Material surfaces can tell a story of the object's past life while semantically conveying how they may be used in the future. In stark contrast, material change may also be perceived negatively as damage, degradation, or contamination: "Plastics cease to be pristine, and become evidently worn, in a particular way. They do not patinate; they gather dirt rather than 'charm,' and then may elicit particularly strong feelings of disgust" (Fisher, 2004, p. 30). Fisher links this negative reaction to a particular material, but more generally people's response to materials is determined by "a societal preoccupation with what an appropriate condition is for certain typologies of material and objects to be in" (Chapman, 2014, p. 141). Unfortunately, a designer cannot simply determine the "appropriate condition" for certain materials in certain contexts to achieve a particular response from the user. Previous studies have shown that attitudes to material aging are highly variable, subjective, and shaped by myriad factors (Bridgens & Lilley, 2017; Lilley et al., 2016, 2019), which include both individual and societal influences, which we will explore later.

6.2 The interaction of material change and material experience

By combining insights from user studies (e.g., Lilley et al., 2016, 2019; Manley et al., 2015) with the existing literature, we have pieced together a tentative framework for understanding the influence of material change on people's experience of the material world which surrounds us. At the heart of the framework is the pioneering work of Nathan Crilly who provided a conceptual framework for understanding consumer response to product visual form (Crilly et al., 2004). Crilly's framework has been expanded to include an understanding of both materials engineering (manufacturing, material properties, and processes of change) and recent work on "materials experience" exemplified by Karana et al. (2015)'s

"Material Driven Design" methodology and wider contributions described in *Materials Experience: fundamentals of materials and design* (Karana et al., 2014). These works provide a basis for understanding the factors which influence a person's response to a particular material in a particular context. We have extended this work by considering the condition of the object, how its condition will change with use, and how this will influence people's response to that object (Fig. 6.2). While "object" is a convenient word to describe the use of a material in a particular context, the framework is equally applicable to material surfaces at any scale, including, for example, furniture, vehicles, the interior and exterior of buildings, and public spaces.

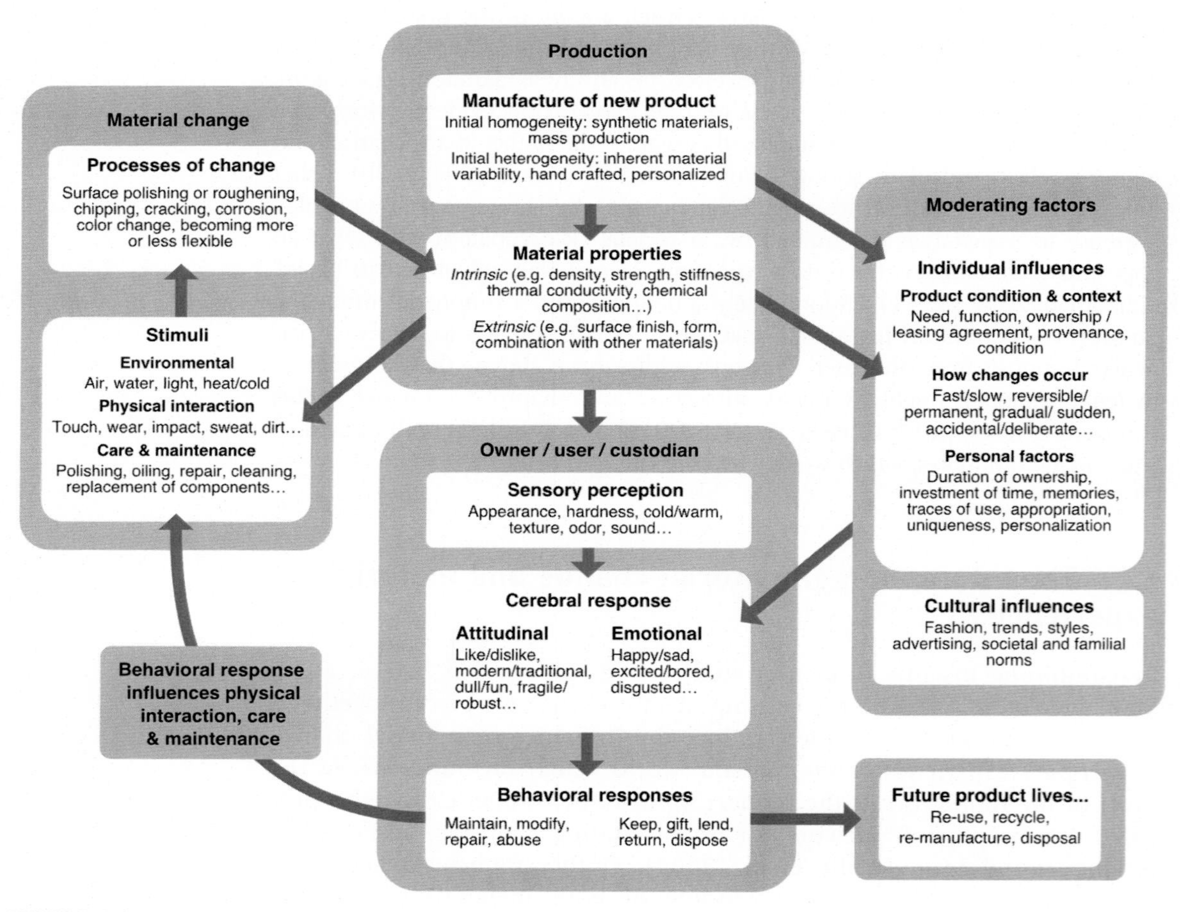

FIGURE 6.2
The interaction of material change and material experience.

6.2.1 Individual influences

The framework shown in Fig. 6.2 identifies numerous "moderating factors," which influence how an individual responds to a particular material, in a particular condition, in a particular context. These are not discrete factors that can be considered individually; they form a complex interdependent web which, when considered in its entirety, describes the elusive difference between graceful aging and degradation.

Change is perceived by comparing the current state of an object or material surface to a remembered previous condition. This observation may sound trivial, but it elucidates two fundamental factors that influence how material change is perceived: the initial condition of the object and the rate at which changes occur.

The initial condition of an object is critical in understanding how material change will be perceived. The "initial condition" might describe a pristine new product that the user fondly remembers removing from its packaging in its "box fresh" state, a well-worn object that was received as a gift or was purchased second hand, or a shared material surface such as a bus seat or the wall of a building, for which the user may not know the original condition and is left to speculate about its past. The "pristine new product" is a generalization of a broad spectrum of materials and manufacturing methods, which result in different reference points for subsequent perception of material change. The vast majority of products are mass-produced using highly regulated man-made materials such as metals and plastics with exquisitely controlled surface finishes, exemplified by the smooth, sleek perfection of portable electronic devices. This uniform, flawless esthetic has insidiously infiltrated into almost every product category, from Ikea's plastic and foil covered particleboard bookshelves to the immaculately smooth painted surfaces of cars. A thin veneer or coating of high-quality material finish is vulnerable to chipping, scratching, and denting, with two problems: first the contrast with the pristine original surface is perceived negatively and second these materials (e.g., unlike solid wood) often cannot easily (or economically) be repaired.

In a backlash against this uniformity, natural materials and "hand-crafted" objects have seen a resurgence of popularity. Unfortunately, they are seen as premium products, and a high price is usually paid for variability and uniqueness. The natural variation inherent in wood, stone, and leather enables unique products to be mass-produced, relying on material variability to distinguish objects of identical shape. Crucially, variability is more often accepted and appreciated in these materials, whereas it is seen as a manufacturing flaw in plastics or metals. These "natural" materials also benefit from surfaces with complex variations in color and texture, making initial changes to the surface less obvious than the first scratch on a polished metal or painted surface. Ceramics and glassware have come full circle from their hand-crafted origins, through mass production of identical pieces, through to "mass customization," which introduces variability into the manufacturing process either by manual intervention (by shaping or decorating) or by adding a random element to an automated manufacturing process. Here the designers are trying to conflate "unique" and "hand-crafted"; time will tell if random variation has the enduring appeal of a craftsman in dialog with a material.

The second fundamental factor relates to the rate of change. Attitudes to changing material surfaces are strongly influenced by *when* changes occur in the cycle of ownership and how quickly or slowly these changes manifest. Sudden changes, for example the first scratch on a pristine manufactured surface, elicit negative reactions due to the jarring change in the product appearance. Gradual changes, however, such as build-up of accumulated dirt or abrasion, allow time for acclimatization and are more likely to be perceived neutrally or positively (Manley et al., 2015). Let's return to our description of the perception of change as the difference between the current condition of a material surface and a remembered previous state. For the first scratch, this stark change is compared to recent memories of the pristine surface and perhaps memories of the moment of purchase. For gradual change, which may only be noticeable over a period of years, memories of the original condition of the object fade and the remembered state alters with time and is never far from the current condition of the object. In extreme cases material change may only be perceptible over decades or centuries. Stone steps and paving slabs wear so gradually that the change is unlikely to be noticed within one lifetime, but worn stones provide a compelling reminder that many generations have lived before us, and that we should design and build not just for ourselves but for future generations.

Materials are not experienced in isolation but in context, forming the visual and tactile interface with the world around us, from toasters to bicycles to buildings. As depicted in Fig. 6.3, just as our response to a material (or more commonly an assemblage of materials) is strongly determined by context (Hekkert & Karana, 2014), our response to material change is equally dependent on context. "Traces of use" on sports equipment or hand tools is often treasured as a record of accomplishment and a reminder of past times, changes to the surface of home furnishings—perhaps the fading of fabrics in sunlight, accumulation of dirt or stains, or a pet cat scratching a sofa—are perceived negatively as damage. Here the distinction seems clear: for a tool or piece of sports equipment the material change occurred as a result of proper use of that equipment and provides a reminder of an

FIGURE 6.3
Diverse forms of material change show the importance of time, context, and esthetics in influencing how material change is experienced. Clockwise from top left: a solid brass doorknob has changed slowly over several decades resulting in a desirable patina on a once homogeneous, pristine material surface; celebrating beautiful repair with Lego (inspired by Jan Vormann); a wooden chair shows intensely personal marks from a baby's teeth; "No Waste" vases by Ikea made from imperfect or damaged glass vases discarded during production, which are melted together and mouth-blown into new, uniquely patterned vases.

enjoyable activity. The changes that have occurred to the home furnishings are accidental; the owner would have preferred that they had not happened. But is this distinction clear, or could the "traces of use" on the home furnishings be widely celebrated as a reminder of good times, with greater consideration of how materials may change during design and material specification?

Public "shared surfaces" provide a particular challenge. Without a feeling of personal ownership, changes to the material surface are more likely to be perceived as damage, accumulation of dirt is more likely to be seen as contamination, and the user is less likely to care for and maintain the object. Returning to our scenario at the outset of the chapter, little care was given to protecting the surface of the cheap, already damaged desk nor to maintaining the interior of the rental car, a common behavior which Bardhi & Eckhardt (2012) observed during their research study.

An object's "intrinsic value" to the individual is also pertinent. Objects can become inextricably woven into our lives as tangible manifestations of past experiences and memories, or discarded when undesirable, superfluous, or outdated. The lifespan of objects is highly variable: from rapidly discarded "fast fashion" to heirloom objects which outlive many generations of owners. Care and maintenance can extend the lifespan of treasured objects indefinitely, or less cared for objects can be rapidly replaced. The greater our regard for an object, the higher our tolerance (or appreciation) of material change. Valued but "worn out" possessions may be judged less critically, their imperfections overlooked or celebrated, their failings forgiven and their surfaces rejuvenated or repaired; "we are less likely to dispose of objects we are emotionally invested in; similarly we are more likely to repair such objects than we are 'insignificant' things" (Harper, 2017, p. 61).

Repair is practiced with the intention to restore function or renew appearance (Zijlema et al., 2017). Depending on its execution and the prevailing esthetic aim, repair activities can irrevocably alter the sensorial qualities of an object and its material constituents; Zijlema et al. (2017) refer to these changes as "traces of repair." These traces can manifest in esthetic alteration (e.g., wood veneer repaired in a contrasting color, ibid) and/or sensorial alteration (a necklace fastening which became inflexible postrepair, ibid). If undertaken with restorative qualities in mind, repair can be seamless and undetectable, rendering an object "as good as new," conversely obvious and unapologetic repair can enhance and give greater esthetic and emotional value making an object "better than new" by virtue of its unique, personalized esthetic. In rare cases obvious repair can increase the monetary value, the prime example being the practice of Kintsugi in which broken ceramics are repaired with gold. Investment of time and effort in repair and maintenance builds an emotional bond with an object. Here the owner takes control of the "processes of change," in contrast to the unavoidable change caused by environmental stimuli or physical interaction which is seen as an external influence. Once material change becomes a conscious, deliberate activity, it is much more likely that the outcome will be seen as a desirable and meaningful patina. However, while skilled maintenance and repair that is material and object appropriate may lead to accrual of value, poor-quality repair may result in a diminished object value (both monetarily and emotionally) unless the act of repair itself is linked to cherished memories or personal achievement. The art of esthetically bold repair, popularized by Sugru who eulogize "the art of beautiful repair," resonates with the celebration of material change. "In a world where almost everything you buy is mass-produced, isn't it nice to have something that is totally unique because of its scars?" (Sugru, 2017). The popularity of Sugru, exemplified through #mysugrufix, demonstrates a growing desire to prolong the life of objects through elevating repair beyond functional or esthetic improvement into a new genre of product esthetic. Yet, the decision not to repair an object is equally revealing. The wish to preserve an item "as it always was" or "as it was remembered" can also be a powerful indication of sustained attachment (Zijlema et al., 2017).

6.2.2 Cultural influences

There are distinct parallels between the preoccupation with youthful perfection and a desire for flawless mass-produced goods: "...our concepts of the ageing of manufactured objects are heavily dependent on their association with our own ageing and with the passing of time measured in human terms" (Scarre, 2016, p. 88). The prevailing esthetic is one of perfection "in every sphere of human life: the body, the style of life, artifacts and their materials" (Rognoli & Karana, 2014, p. 147). However, "just like living beings mature and get older, so too artifacts degrade, and their surfaces show signs of aging" (Rognoli & Karana, 2014, p. 150). Are negative perceptions of aging possessions driven by deeper systemic cultural, esthetic, and philosophical responses to human aging? Could an understanding of the relationship between people's responses to the esthetic aging of objects and their own aging inform design strategies to combat esthetic obsolescence? (Bridgens et al., 2019).

The pressure to conform to "plasticized" beauty belies the value of irregularity, imperfection, and authenticity inherent in the philosophies of wabi-sabi which prizes "the beauty of faded, eroded, oxidized, scratched, intimate, rough, earthy, vanishing, elusive, ephemeral things" (Salvia et al., 2010, p. 1580) and *kintsugi*—in which "the beauty of that which was broken and visibly mended" is celebrated (Buetow & Wallis, 2019). The concealment and eradication of wrinkles, lines, and visible scars erases "traces" of the past; the inscription and repository of time and memories represented within the skin's surface. These "ephemeral patterns" are "signs of ... an ongoing engagement with life." Perhaps they should not be hidden or obscured but celebrated, valued, and appreciated (Buetow & Wallis, 2019), in the same way that "traces of use" on a material surface can reveal its narrative "in the same way that the scar on a lover's leg or the wrinkles around his eyes appear beautiful ... because they contain the stories about the time he fell off his bicycle ... aesthetic decay possesses qualities that make an object more valuable" (Harper, 2017, p. 77).

To achieve such a mindset requires a shift in cultural ideals and changes to current notions of beauty and perfection. It might be argued that this is beyond the reach of designers' influence. But as the creators of the material world, architects of commoditized communities and purveyors of dreams and ideals are not designers perfectly placed to challenge such embedded attitudes? By cultivating greater tolerance of material change and aging in our objects, might we affect a wider cultural transition toward an acceptance (or even celebration) of our aging selves?

6.3 Material change as a design strategy

In *Materials Experience: fundamentals of materials and design* (Karana et al., 2014), the promise of graceful aging as a design strategy to combat early obsolescence is highlighted. In this chapter we ask, "what if 'material change' was an overt design strategy?" What if designers could carefully orchestrate material change through the life of an object, could determine or influence how the look and feel of objects would change in response to use and better understand and control how these changes would alter people's attitudes toward owned or shared objects, and the object's treatment, lifespan, and future? How might an increased appreciation of changing material surfaces alter our experiences of, and relationships with, objects? We propose that if designers harnessed material change, they could actively choose to halt, slow, hasten, or modify material transformations to manipulate how a material looks, feels, smells, and sounds as it ages. Examples exist that utilize material change as a communicative strategy, from syringes which change color to signal when they are no longer sterile and packaging which indicates when food is still safe to eat, or to address functional needs such as cracks in buildings which can close on their own, car bodies that can recover their original shiny appearance by themselves and mobile phones which can detect and fix cracks in their screens. However, little consideration has been given to how material change may direct, inform, and create user experiences through shifting attitudes and behavior. Designs that embody material change as a strategy, as seen in Fig. 6.4, are few in number and are often conceptual prototypes.

What might material change as a design strategy look like?
How might it be manifested in the design of objects and environments?

It is not easy to predict how people will respond to material change, but we argue that designers can use materials more effectively to ensure their intentions for evoking particular attitudes and behaviors are realized when interacting with designed objects or environments and the materials they embody. Through increased focus on user behavior, designers can purposefully shape the way users interact with products (and by extension materials) to leverage more sustainable use patterns. We call this approach Design for Sustainable Behavior (Lilley et al., 2017). When considering the suite of strategies available (see Lilley et al., 2017 for a comprehensive overview), the approaches that are most suitable for designing with material change are the provision of olfactory, aural, or esthetic feedback and the use of the physical, sensorial, or semantic characteristics of the materials or their affordances (explicit potential actions) and constraints (explicit potential limitations), an approach we call "behavior steering" (ibid). Let's consider two scenarios that place these approaches in a materials context:

FIGURE 6.4

Material change as design strategy. Clockwise from top left: Underfull table cloth https://design-milk.com/underfull-table-cloth-by-kristine-bjaadal/; ENDURE concept https://youtu.be/iEQzNMawTjg; Stain TeaCup http://www.bethanlaurawood.com/work/stain/.

In the context of public "shared surfaces" the default design strategy is to use highly durable materials that will endure constant public use with minimal change, resulting in gleaming, impersonal, echoing public spaces lined with glass, stainless steel, and polished granite. But these shared surfaces can provide opportunities to incorporate material change in design, by guiding and recording the behavior of large numbers of people. The result is "desire paths" worn through landscaped areas, old stone steps worn from hundreds of years of use, and on ticket machines the most popular destinations clearly signposted by the wear and polish on certain buttons (before the advent of touch-screen interfaces). Deliberate use of materials that wear, polish, and change in response to use has scope to transform our public spaces into places where people feel that the space is shaped by them, and the environment is richly laden with semantic clues about how to navigate an otherwise unfamiliar realm.

And within the context of the home:

> Scenario 2: "The shiny red surfaces of your polypropylene dining chairs reflect the intense summer sun which floods your dining room. Each chair is uniquely patinated in response to variations in UV exposure and their position in your home. You marvel at their beauty, so much more at home than they were when you unpacked six identical, plain red chairs last year. You sit, the curves of the seat folding around you, melding its contours to yours. Your hand rests on the simple white, damask tablecloth which reveals the rich tapestry of last night's dinner with friends. Carla's wine spilt during a lively Brexit debate immortalized in the joyous flight of butterflies dancing across the surface. Your eyes rest on your children's highchair and a burst of images of Jake flit through your mind. The initial gleam of varnish has faded, the bite marks on the backrest a bittersweet reminder of nights spent comforting your teething baby. You reach for your teacup and smile. Time and tannin have stained its surface. You feel the warmth of the cup in your hands and ponder the passing of time. Life is fleeting but the familiar 'traces of use' remain."

6.4 Conclusion

Materials change. We experience the world around us by interacting with constantly changing material surfaces. Our responses and attitudes to these material transformations are as complex as the changes themselves. There is an expectation that the engineered surfaces of "man-made" materials should somehow be impervious to change, coupled to a widely held belief that "natural" materials age well. In truth, neither perception is entirely accurate as "the social values affixed to the aging of material surfaces are intensely complex" (Chapman, 2014, p. 141). Our experience of material surfaces in a state of flux varies from disappointment and disgust to cherishing objects to which we feel a deep emotional connection.

We see enormous potential for designers to influence material experience beyond the moment of purchase, by considering material change as a fundamental part of the design process. But manipulating or orchestrating material change is not straightforward. To leverage this approach, designers must be cognizant of the capabilities of different materials and the stimuli that result in surface changes, how people will interact with the materials, and how they will respond to the resulting changes. In addition to this knowledge and understanding, designers must also develop a clear and deliberate strategy for material change, interrogating their own intent (which behaviors do they wish to encourage/discourage? what experiences are they aiming to create?) as well as the ethical implications of intended and unintended behavioral outcomes (Lilley & Wilson, 2013). Those among the profession practicing responsibly should seek to use material change as a force for good, creating virtuous loops (care, maintenance, and prolonged life) rather than destructive behaviors (premature disposal, misuse, and damage).

References

Bardhi, F., & Eckhardt, G. M. (2012). Access-based consumption: The case of car sharing. *Journal of Consumer Research, 39*(4), 881–898.

Bridgens, B., & Lilley, D. (2017). Design for next... year. The challenge of designing for material change. *The Design Journal, 20*(sup1), S160–S171.

Bridgens, B., Lilley, D., Zeilig, H., & Searing, C. (2019). Skin deep. Perceptions of human and material ageing and opportunities for design. *The Design Journal, 22*(Suppl. 1), 2251–2255.

Buetow, S., & Wallis, K. (2019). The beauty in perfect imperfection. *Journal of Medical Humanities, 40*, 389–394.

Chapman, J. (2014). Meaningful stuff: Towards longer lasting products. In E. Karana, O. Pedgley, & V. Rognoli (Eds.), *Materials experience: Fundamentals of materials and design* (pp. 135–143). Oxford: Butterworth-Heinemann.

Crilly, N., Moultrie, J., & Clarkson, P. J. (2004). Seeing things: Consumer response to the visual domain in product design. *Design Studies, 25*(6), 547–577.

Fisher, T. H. (2004). What we touch, touches us: Materials, affects, and affordances. *Design Issues, 20*(4), 20–31.

Giaccardi, E., Karana, E., Robbins, H., & D'Olivo, P. (2014). Growing traces on objects of daily use: A product design perspective for HCI. In *Proceedings of the 2014 conference on designing interactive systems*.

Harper, K. H. (2017). *Aesthetic sustainability: Product design and sustainable usage*. Routledge.

Hekkert, P., & Karana, E. (2014). Designing material experience. In E. Karana, O. Pedgley, & V. Rognoli (Eds.), *Materials experience: Fundamentals of materials and design* (pp. 3–11). Oxford: Butterworth-Heinemann.

Karana, E., Barati, B., Rognoli, V., & Zeeuw Van Der Laan, A. (2015). Material driven design (MDD): A method to design for material experiences. *International Journal of Design, 19*(2).

Karana, E., Pedgley, O., & Rognoli, V. (2014). *Materials experience: Fundamentals of materials and design*. Oxford: Butterworth-Heinemann.

Lilley, D., Bridgens, B., Davies, A., & Holstov, A. (2019). Ageing (dis) gracefully: Enabling designers to understand material change. *Journal of Cleaner Production, 220*, 417–430.

Lilley, D., Smalley, G., Bridgens, B., Wilson, G. T., & Balasundaram, K. (2016). Cosmetic obsolescence? User perceptions of new and artificially aged materials. *Materials & Design, 101*, 355–365.

Lilley, D., & Wilson, G. (2013). Integrating ethics into design for sustainable behaviour. *Journal of Design Research, 11*(3), 278–299.

Lilley, D., Wilson, G., Bhamra, T., Hanratty, M., & Tang, T. (2017). Design interventions for sustainable behaviour. *In Design for behaviour change* (pp. 40–57). Routledge.

Manley, A., Lilley, D., & Hurn, K. (2015). Cosmetic wear and affective responses in digital products: Towards an understanding of what types of cosmetic wear cause what types of attitudinal responses from smartphone users. In *Proceedings of product lifetimes and the environment (PLATE)*, Nottingham Trent University.

Nobels, E., Ostuzzi, F., Levi, M., Rognoli, V., & Detand, J. (2015). Materials, time and emotion: how materials change in time? In *EKSIG 2015 TANGIBLE MEANS experiential knowledge through materials*.

Odom, W., & Pierce, J. (2009). Improving with age: designing enduring interactive products. *In CHI'09 extended abstracts on human factors in computing systems* (pp. 3793–3798). Boston: ACM.

Pye, D. (1968). *The nature and art of workmanship*. London: A&C Black Publishers.

Rognoli, V., & Karana, E. (2014). Towards a new materials aesthetic based on imperfection and graceful ageing. In E. Karana, O. Pedgley, & V. Rognoli (Eds.), *Materials experience: fundamentals of materials and design* (pp. 145–154). Oxford: Butterworth-Heinemann.

Salvia, G., Ostuzzi, F., Rognoli, V., & Levi, M. (2010). *The* value of imperfection in sustainable design. In Sustainability in Design: Now (pp. 1573–1589).

Scarre, G. (2016). The ageing of people and of things. *In The Palgrave handbook of the philosophy of aging* (pp. 87–99). Springer.

Sugru. *The art of beautiful repairs*. (2017). <https://sugru.com/blog/the-art-of-beautiful-repair> Accessed 15.09.20.

Zijlema, A., van den Hoven, E., & Eggen, J. (2017). Preserving objects, preserving memories: Repair professionals and object owners on the relation between traces on personal possessions and memories. *Delft University of Technology*, *8*, 10.

PROJECTS

Around The Corner: Recent and Ongoing Research in Materials and Design

PROJECT 1

Design touch matters: bending and stretching the potentials of smart material composites

Bahareh Barati
Faculty of Industrial Design Engineering, Delft University of Technology, Delft, Netherlands

Supervisors: Prof. Dr. Elvin Karana and Prof. Dr. Paul Hekkert

My PhD thesis on the topic of understanding the experiential qualities of smart material composites in their early development stages was developed through the compilation of three peer-reviewed journal papers and one ACM conference paper on human factors in computing systems (CHI' 18). Across the papers, I describe how I employed various design research methods, including design projects, research through design (in developing a toolkit to prototype the dynamic and performative qualities of these materials), and semistructured interviews.

The first exploratory study in my research focused on the situation of designing with an underdeveloped smart material composite, as experienced in the LTM (Light. Touch. Matters) project. Observational studies and interviews with LTM designers suggested that dynamic and performative qualities of smart material composites are hardly able to be represented through a listing of their physical and functional properties. In response, as well as developing and testing a hybrid prototyping toolkit (combining physical and digital prototyping tools), the research took a fundamental turn to investigate the role of material tinkering and fabrication in unlocking novel material potentials.

Consequently, I initiated several material-driven design projects, focusing on electroluminescent (EL) material and supported designers in directly working with the physical material, rather than considering just the specifications of its properties. The material-driven design processes provided empirical evidence that creativity in materials and design often involves discovering novel material affordances that are not anticipated or intended without the benefit of tinkering and physical processing of the materials.

In addressing the main research question (i.e., how do designers unlock the potentials of underdeveloped smart material composites?), the notion of "affordances as materials potential" was introduced and used to discuss the creative materiality contribution of designers that goes beyond product novelty. An immediate benefit of the affordance concept for capturing the "potential" of a material was that it enables descriptions of the material in terms of process abilities. By contrasting the two design situations (i.e., departing from an actual material, vs descriptions of its properties), I identified and laid bare the limitations of product-oriented approaches that characterize underutilized smart material composites as "black-boxes," in contrast to investing in understanding and exploring their unique affordances.

Barati, B., & Karana, E. (2019). Affordances as materials potential: What design can do for materials development. *International Journal of Design, 13*(3), 105–123.

Barati, B., Giaccardi, E., & Karana, E. (2018). The making of performativity in designing [with] smart material composites. In *Proceedings of the 2018 CHI conference on human factors in computing systems (CHI'18)* (pp. 5:1–5:11). Montreal: ACM.

Materials Experience 2. DOI: https://doi.org/10.1016/B978-0-12-819244-3.00030-2

Why is the research needed?

The research sought to understand what and how "design" (as a profession, as a way of thinking, as an activity, etc.) can contribute to the early stages of developing new smart material composites. The EU-funded project, Light. Touch. Matters (LTM, 2013–17) and its organization set a departure point and provided the context for further investigations.

Which aspects of "materials experience" does the research valorize?

The research elaborates on the performative level of materials experience in designing [with] smart material composites.

How will the research impact on designers?

Designers' relationships with materials are challenged —the research articulates designers' creative contribution to upstream collaborative materials development, which directly questions the taken-for-granted separated roles of designers and scientists to, respectively, conceptualize products and develop materials.

What outcomes have been achieved or are foreseen?

The research provided a theoretical framework (aiming to expand the definitions of materials' potential), a design toolkit (aiming to represent and communicate the experiential qualities of an underdeveloped smart material), a method (for design processes that depart from a material), and a material-driven design strategy (for creating performative smart material composites).

What is the next big challenge for the research area?

The challenge lies in educating and supporting the next generation of designers who can effectively collaborate with scientists and engineers in materials development.

Which publications most inspired or informed the research?

Glaveanu, V.P. (2014). *Distributed creativity: Thinking outside the box of the creative individual*. Springer Science & Business Media.

Bergström, J., Clark, B., Frigo, A., Mazé, R., Redström, J., & Vallgårda, A. (2010). Becoming materials: Material forms and forms of practice. *Digital Creativity*, *21*(3), 155–172.

Miodownik, M. (2007). Toward designing new sensoaesthetic materials. *Pure and Applied Chemistry*, *79* (10), 1635–1641.

Gaver, W.W. (1996). Situating action II: Affordances for interaction: The social is material for design. *Ecological Psychology*, *8*(2), 111–129.

Acknowledgments

This work was part of the European Union (EU) project Light Touch Matters and received funding from the EU's Seventh Framework Program (FP7/2007–2013) under grant agreement number 310311.

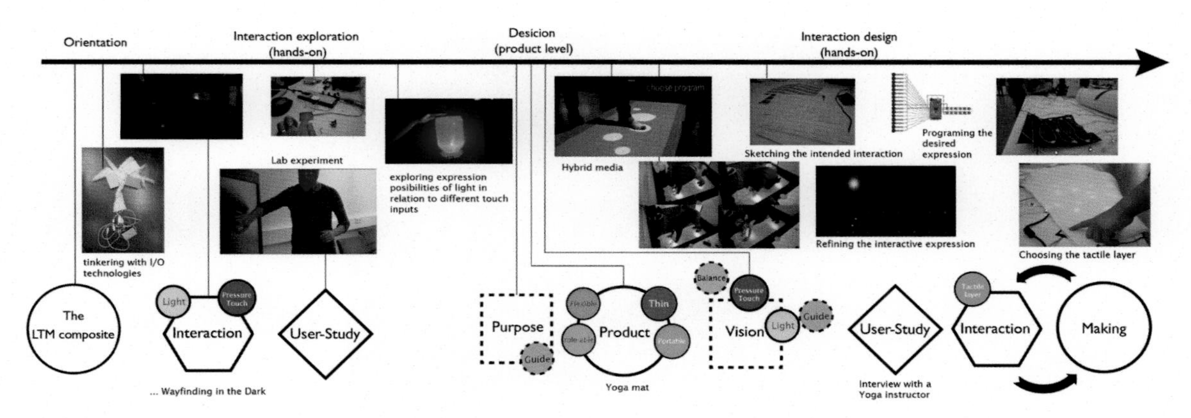

FIGURE 1

Analysis of the activities and the prototypes across a design process departing from physical and functional descriptions of LTM materials. *The design project was carried out by master's students of Interactive Technology Design course, TU Delft, 2014.*

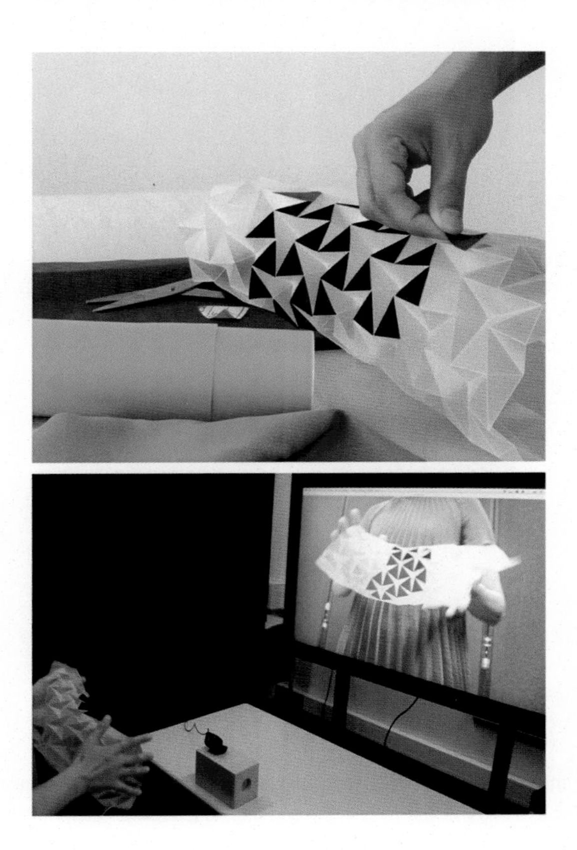

FIGURE 2

Chroma key prototyping toolkit proposed to support exploration and communication of experiential qualities of an underdeveloped smart material composite: preparation of the physical mock-up by adding velvety stickers (top) and the appearance of the real-time Chroma-key effect on the screen (bottom). *Design and photo credits by the author.*

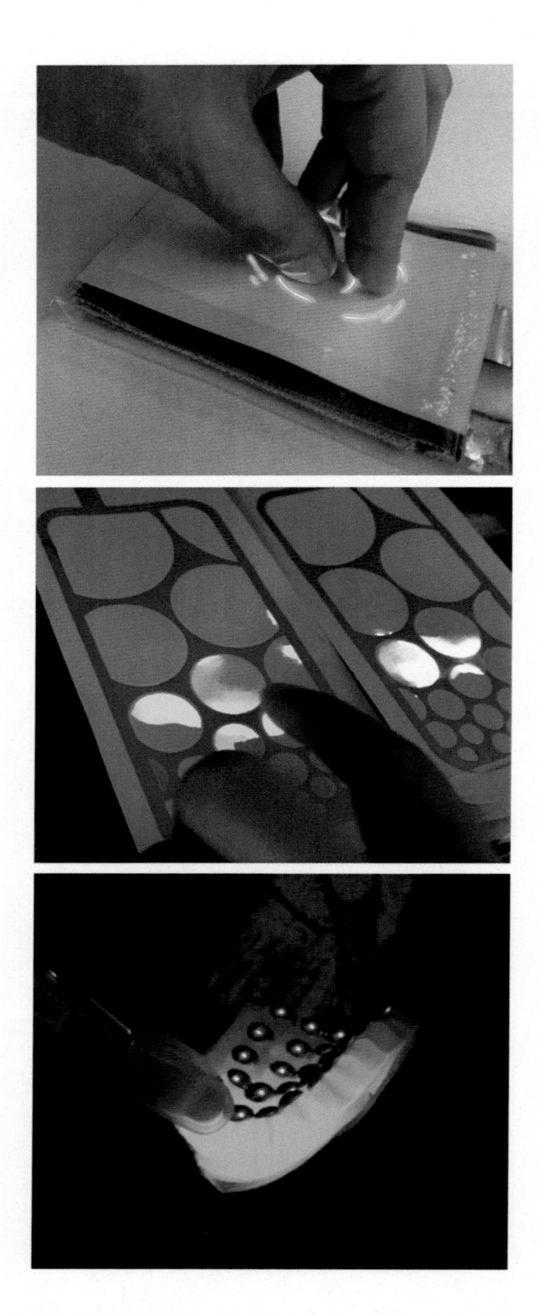

FIGURE 3
Designers encourage new actions through alterations to the basic ingredient, structure, form, and elements of computation when processing electroluminescent samples: pressing and pinching to see the hidden texture (top), brushing the wet surface with the fingers (center), squeezing and kneading to spread the light (bottom). *Design credits by Stan Claus (top and middle) and the student of Material Driven Design course (bottom), TU Delft, 2016–2017.*

FIGURE 4

The final concept of a material-driven design process pushes for a novel experience of rain, using the custom-made water-activated EL print (top). *Master's graduation project by Stan Claus, TU Delft, 2016. Photo credits: Wiersma Brothers.*

PROJECT 2

Design for hybrid material systems: a material augmentation framework for meaningful experiences

Stefano Parisi
Design Department, Politecnico di Milano, Milan, Italy

Supervisor: Assoc. Prof. Dr. Valentina Rognoli

My research focuses on Hybrid Material Systems, defined as material-based systems combining inactive materials, stimuli-responsive smart materials, and embedded sensing, computing, and actuating micro-technologies. They arise as emerging materials and technologies to be introduced as tangible interfaces in a wide range of sectors for esthetic enjoyment and functional enhancement.

The research adopts a speculative approach due to the presence of current technological limitations that will eventually be overcome through miniaturization and advancement. I aim to identify meaningful and purposeful ways for designing with and for Hybrid Material Systems. Indeed, the integration of such materials into everyday design practice and into applications that foster people's appreciation and acceptance implies complex and novel experiences that designers need to comprehend and master.

What are the meaningful experiential patterns Hybrid Material Systems enable and imply for users and designers? What are the relations and entanglements between their material qualities and their interactive behaviors? To obtain answers, my research uses a mixed-method design, collecting mainly qualitative data through a set of studies based on observations of projects, questionnaires, and research through design using Do-It-Yourself (DIY) materials and material tinkering.

I have progressively organized the results and findings from each study into a framework composed of tools, methods, approaches, procedures, and guidelines (Fig. 1). These are grounded in material-driven design, systemic design, speculative design, metaphors and analogies, and biomimicry. Hybrid Material Systems based on alternative bio-based materials with embedded electronics and smart components are revealing excellent potential as a raw material for tinkering and iterative experimentation, soft robotics, and ephemeral interface applications, while also responding to the demand for more sustainable materials (Fig. 2).

Parisi, S., Rognoli, V. (2021). Design for ICS Materials: the development of tools and methods for the inspiration and ideation phases. In: V. Rognoli, V. Ferraro (Eds), ICS Materials: interactive, connected, and smart materials, 203-217. Design International series. Franco Angeli. http://ojs.francoangeli.it/_omp/index.php/oa/catalog/book/641

Parisi, S., Rognoli, V., Spallazzo, D., & Petrelli, D. (2018). ICS materials: Towards a re-interpretation of material qualities through interactive, connected, and smart materials. In Proceedings of DRS 2018 international conference of the design research society volume 4 (pp. 1747–1761). Limerick: University of Limerick.

Why is the research needed?

The research aims to support designers and design students in understanding, conceptualizing, and designing

Materials Experience 2. DOI: https://doi.org/10.1016/B978-0-12-819244-3.00014-4

with and for complex materials systems with dynamic, interactive, and smart behaviors—named as "Hybrid Material Systems"—for meaningful experiences that foster people's appreciation.

Which aspects of "materials experience" does the research valorize?

The research unfolds and discusses significant and recurrent experiential patterns enabled by materials with interactive qualities. It articulates the relations between their static and temporal forms, emphasizing the fluctuating nature of materials experience.

How will the research impact designers?

Designers and design students can benefit from novel and systematically organized tools, methods, approaches, and notions when designing with and for complex material systems with dynamic behaviors, to more easily comprehend and enhance meaningful experiences.

What outcomes have been achieved or are foreseen?

The research proposes an ontology and taxonomy of Hybrid Material Systems and identifies a framework to design with and for them. This is implemented into a set of tools, methods, procedures, guidelines, and learning contents, which have been applied and tested mainly in design workshops with students, delivering concepts and prototypes (Fig. 3 and 4).

What is the next big challenge for the research area?

Sustainability represents a challenge and a topic to explore when dealing with Hybrid Material Systems, in particular, regarding the use of organic materials or the implementation of bio-sensors and bio-actuators to replace electronics.

Which publications most inspired or informed the research?

Giaccardi, E., & Karana, E. (2015). Foundations of materials experience: An approach for HCI. In *Proceedings of the 33rd annual ACM conference on human factors in computing systems CHI'15* (pp. 2447–2456). New York: Association for Computing Machinery.

Razzaque, M. A., Delaney, K., & Dobson, S. (2013). Augmented materials: Spatially embodied sensor networks. *International Journal of Communication Networks and Distributed Systems, 11*(4), 453–477.

Vallgårda, A. (2009). *Computational composites: Understanding the materiality of computational technology* [PhD thesis, The IT University of Copenhagen, Copenhagen].

Ritter, A. (2006). *Smart materials in architecture, interior architecture and design*. Basel: Birkhauser.

Acknowledgments

The research has been partially funded by Fondazione F.lli Confalonieri. Parts of the research have been carried out in cooperation with the projects ICS_Materials (http://www.icsmaterials.polimi.it) and InDATA (http://www.indata.polimi.it) at Politecnico di Milano, with the Digital Materiality Lab at Sheffield Hallam University and the Institute for Material Design at Offenbach School of Arts and Design.

FIGURE 1

Applying and testing tools and methods in the workshop “NautICS Materials” with Master in Yacht Design students at POLI.design, Milano, with Arianna Bionda and Andrea Ratti, 2018. © *Author.*

FIGURE 2

Tools and guidelines were applied in the production of light-emitting bioplastic samples due to the inclusion of light-emitting diodes (LEDs), in the workshop "Coded Bodies" by Giulia Tomasello at the Politecnico di Milano, School of Design, Design for the Fashion System Studio (Prof. Paola Bertola), 2019. Samples realized by Mara Iannoni, Shiva Jabari, Aleksandra Obradovid, and Yang Xiaoxuan. © *Author.*

FIGURE 3

Responsive light-emitting visor from "Data < > Materials" Hackathon with the InDATA project team (with Ilaria Mariani, Patrizia Bolzan, Mila Stepanovic, and Laura Varisco), at the Politecnico di Milano, School of Design, 2019. Protoype and concept realized by Anna Vezzali, Daniele Carlini, Davide Franci, Ivan Oda, and Pietro Lora. © *Author.*

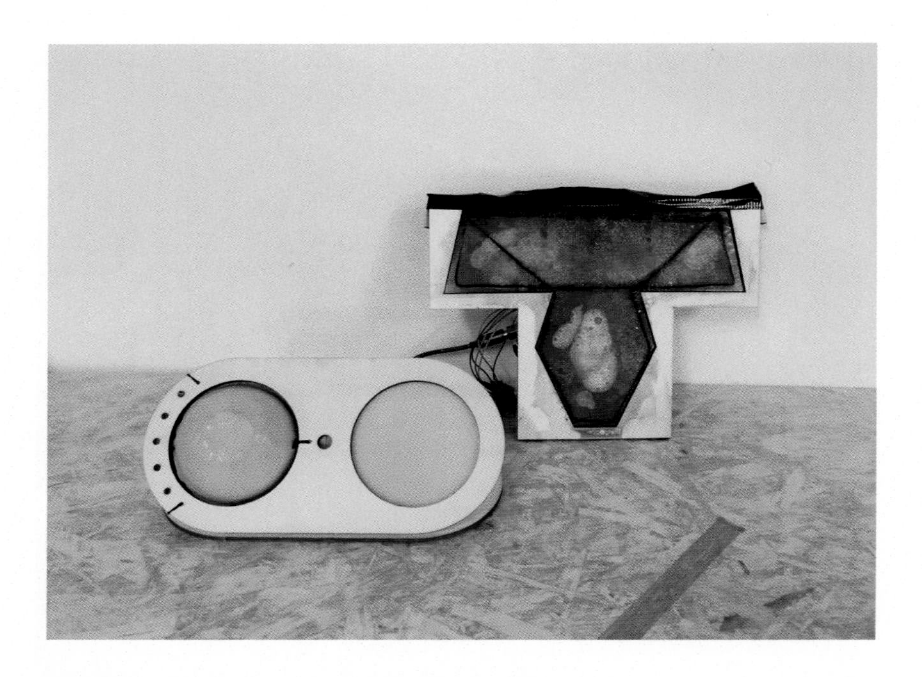

FIGURE 4

Touch-sensitive vibrating and light-emitting bio-skin and responsive light-emitting mask from "Data < > Materials" Hackathon with the InDATA project team (with Ilaria Mariani, Patrizia Bolzan, Mila Stepanovic, and Laura Varisco) at the Politecnico di Milano, School of Design, 2019. Prototypes and concepts realized by Emanuele Belà, Marjia Nikolic, Li Chen, Aurelie Glaser, Davide Minighin, Leonardo Saletta, Andrea Torrone, and Elena Ukulova. © *Author.*

An investigation of the esthetics and technologies of photochromic textiles

Dilusha Rajapakse
School of Arts and Design, Nottingham Trent University, Nottingham, United Kingdom

Supervisors: Prof. Amanda Briggs-Goode and Prof. Tilak Dias

The ability to transform the expression of a material surface through changes in colors and decorative patterns can be an effective design feature, not only because of the huge potential for creating a dynamic effect but also with respect to esthetic and emotional pleasure. If these transformations repeatedly and instantly occur as a response to external environmental stimuli and then gradually disappear without leaving any traces, they can be visually stimulating and esthetically pleasing for observers. The concept behind my practice-based design research project was to invent and integrate such transformable dynamic esthetic experiences onto textile surfaces with the application of smart photochromic materials.

By adopting a "research through design" approach, I examined the creative textile design potential of photochromic materials (water-based photochromic inks) in relation to the areas of (1) associated visual characteristics, (2) flat-bed screen printing as a method of the application process, and (3) solar UV (ultraviolet) radiation, artificial UV light sources and SMD (surface mounted device) UV LEDs (light-emitting diodes) as potential activation methods for the excitation of photochromic colorants on textile surfaces. My design thinking and print design practice were employed as leading methods of inquiry, generating textual, numerical, and visual data directly through experimentation with materials, print processes, activation methods, and related technologies. During the experimental stages of the research, specific visual data capturing procedures were consistently executed, and all stages of the research process were complemented with critical reflective practice. This enabled me to obtain an in-depth insight into the behavior of the materials and translate my experiential learning into documented explicit knowledge.

The exploration of the visual characteristics, experimentation of flat-bed screen printing processes, and different activation technologies revealed a number of decisive parameters that could be executed to obtain a new level of dynamic photochromic esthetic outcomes on textile surfaces.

Rajapakse, D. (2019). An investigation of the activation of multi-colour changing photochromic textiles. *Journal of Textiles and Engineer, 26*(114), 196–208.

Rajapakse, D., Briggs-Goode, A., & Dias, T. (2015). Electronically controllable colour changing textile design. In V. Popovic, A. Blackler, N. Nimkulrat, B. Kraal, Y. Nagai & D. Luh (Eds.), *Proceedings of IASDR 2015 interplay* (pp. 1743–1759). Brisbane: The International Association of Societies of Design Research.

Why is the research needed?

This research attempts to investigate the unexploited material characteristics and related design potential of photochromic colorants on textile surfaces to create new material experiences. The research further highlights the possibilities of creatively integrating

Materials Experience 2. DOI: https://doi.org/10.1016/B978-0-12-819244-3.00019-3

multicolor- and pattern-changing photochromic effects onto textiles without compromising the core textile properties of softness, flexibility, washability, drapability, and comfort.

Which aspects of "materials experience" does the research valorize?

The research exploits the functional and expressive/animated color-changing characteristics of photochromic materials which can add to the visual experience in daylight conditions.

How will the research impact on designers?

The theoretical and technical knowledge generated in this research can guide future design practitioners to take informed approaches to the handling of photochromic materials on textile surfaces.

What outcomes have been achieved or are foreseen?

A design-centered approach has been articulated that introduces a number of generalizable decisive parameters associated with design, application, and activation of dynamic photochromic effects on textile surfaces. Furthermore, a unique collection of expression-changing photochromic textile samples/prototypes has been produced, each capable of instantly transforming from one expression to another when exposed to UV radiation.

What is the next big challenge for the research area?

The identification of methodological and technical approaches for the quantification of multicolor-changing photochromic effects will be an important next step.

Which publications most inspired or informed the research?

Stylios, G., & Chen, M. (2016). Psychotextiles and their interaction with the human brain. In V. Koncar (Ed.), *Smart textiles and their application* (1st ed.) (pp. 197–239). Cambridge: Woodhead Publishing Ltd.

Viková, M., Christie, R., & Vik, M. (2014). A unique device for measurement of photochromic textiles. *Research Journal of Textile and Apparel, 18*(1), 6–14.

Little, A., & Christie, R. (2011). Textile applications of photochromic dyes. Part 3: factors affecting the technical performance of textiles screen-printed with commercial photochromic dyes. *Coloration Technology, 127*(5), 275–281.

Worbin, L. (2010). *Designing dynamic textile patterns* [Doctoral thesis, University of Borås, Borås]. <http://urn.kb.se/resolve?urn = urn:nbn:se:hb:diva-3552> Accessed 08.06.20.

Acknowledgments

This research is fully funded by the Vice-Chancellor's PhD Scholarship provided by the Nottingham Trent University.

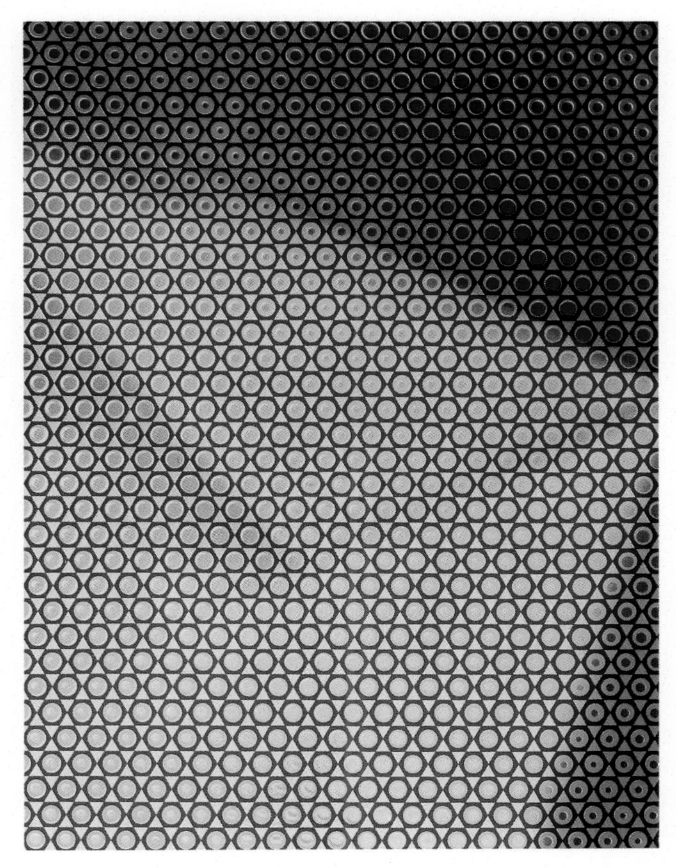

FIGURE 1
Upon exposure to UV, the printed geometrical design morphs reversibly between complementary color expressions. © *Author.*

FIGURE 2
Under UV light, the textile surface changes from a monochromatic design to a multicolored floral design. © *Author.*

FIGURE 3
Under UV light, the textile surface changes from a geometrical pattern to a multicolored floral design. © *Author.*

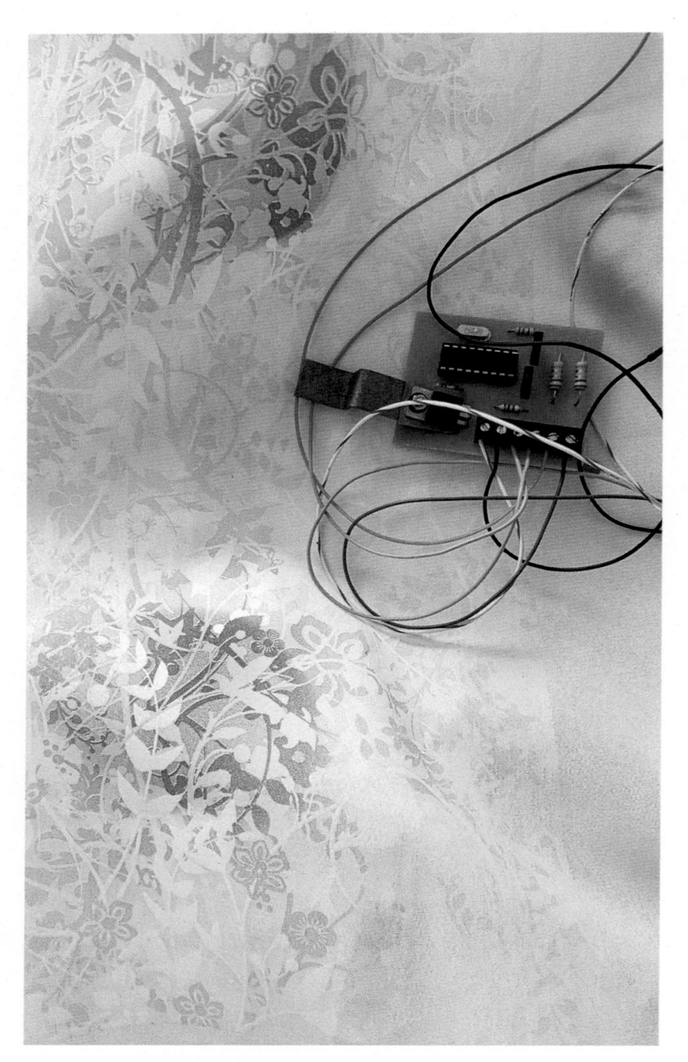

FIGURE 4
Electronically activating and controlling the dynamic photochromic effect on textile surfaces. © *Author.*

Reflective weaving practice in smart textile material development process

Emmi Anna Maria Pouta
Department of Design/Department of Communications and Networking, Aalto University, Espoo, Finland

Supervisors: Assoc. Prof. Kirsi Niinimäki and Asst. Prof. Yu Xiao

Jussi Ville Mikkonen
Department of Electronics and Communications Engineering, Tampere University of Technology, Tampere, Finland

Supervisor: Prof. Dr. Jukka Vanhala

Our work is research-through-design, examining how electrical and sensorial properties can be merged in complex woven structures to expand the experiential qualities of woven smart materials in user interface design. The main methods combine prototyping practices from electronics and textile design, emphasizing the first author's reflective weaving practice and user testing in a controlled environment. The analysis of the research is grounded on transdisciplinary evaluation—electronics engineering and textile design. We discuss prototypical material examples through the lens of reflective weaving practice, in particular, woven e-Textile design that considers a synthesis of different interaction design, electronics engineering, and textile design aspects. We have explored the use of different weaving techniques, such as jacquard weaving, fil coupé-technique, and multilayer double cloth in seven design experiments, consisting of multilayer circuitry and component integration, touch-sensitive sensor structures, and a fully integrated woven user interface. The seven design experiments form a continuum, evolving from an electronics-centered to material-centered process.

In the first two design experiments, we explored digital textiles by integrating electronics components to a weave from a traditional electronics circuit perspective, providing a basis for electronics-led discussion. The first experiment integrated sensors that react to a moving magnet, while the second involved a development of a weaveable component yarn containing a microprocessor. These examples allowed an examination of the separation of the material and the material interaction behavior, highlighted by the electrical material qualities during an interaction.

Design experiments 3–6 focused on developing touch-sensitive structures by utilizing different fibers, weaves, and visual elements. The focus shifted from a thorough investigation of the electrical potential of multilayer structures to the sensorial properties of woven e-Textiles. The seventh design experiment was an interactive hand-puppet, situated in a user-centered smart textile development process. The user experience and preferred interactions were a guideline for the e-Textile design process, dominating the development of the technical construction.

Our woven samples informed the textile behavior and weaving experience, providing cases with which to analyze how the material guided the weaving process. Through analyzing the reflective notes and materials in these processes, we examine how textile thinking and the materiality of interaction intertwine with Schön's concept of "reflective practice."

Materials Experience 2. DOI: https://doi.org/10.1016/B978-0-12-819244-3.00018-1

Pouta, E., & Mikkonen, J. (2019). Hand puppet as means for eTextile synthesis. In *Proceedings of the thirteenth international conference on tangible, embedded, and embodied interaction (TEI '19)* (pp. 415–421). Tempe: ACM Press. doi:10.1145/3294109.3300987.

Mikkonen, J., & Pouta, E. (2016). Flexible wire-component for weaving electronic textiles. In *Proceedings of IEEE 66th electronic components and technology conference (ECTC)* (pp. 1656–1663). Las Vegas: IEEE. doi:10.1109/ECTC.2016.180.

Why is the research needed?

The research focuses on investigating how materiality of interaction—a key part of materials experience—and textile thinking can be comprehensively applied to woven e-Textile development, through a reflective weaving process, and how the design and development process of a smart textile artifact can be synthesized in woven eTextile design.

Which aspects of "materials experience" does the research valorize?

The work extends the understanding of reflective smart material development, more specifically on how electrical and sensorial properties of the materials translate into functional and experiential qualities of user interfaces.

How will the research impact on designers?

Woven multilayer e-Textiles is an uncharted domain, requiring a deep understanding of woven structures and material behavior in an interdisciplinary context. Our research provides guidelines on how to navigate the area.

What outcomes have been achieved or are foreseen?

First, hand-woven e-Textile samples have been created. The samples aid understanding of how sensorial and electrical properties of e-Textile materials intertwine in complex woven textiles. Second, a reflective weaving practice for smart textiles has been developed, comprising a method for guiding multilayer weaving of smart materials through an analytical process.

What is the next big challenge for the research area?

For the future, a better understanding is needed of what a textile-specific interaction can be, with the ability to develop suitable e-Textile interfaces in an interdisciplinary team.

Which publications most inspired or informed the research?

Devendorf, L., & Di Lauro, C. (2019). Adapting double weaving and yarn plying techniques for smart textiles applications. In *Proceedings of the thirteenth international conference on tangible, embedded, and embodied interaction* (pp. 77–85). Tempe: ACM.

Gowrishankar, R., Bredies, K., & Ylirisku, S. (2017). A strategy for material-specific e-Textile interaction design. In S. Schneegass & O. Amft (Eds.), *Smart textiles: Fundamentals, design, and interaction* (pp. 233–257). Cham: Springer International Publishing.

Poupyrev, I., Gong, N., Fukuhara, S., Karagozler, M., Schwesig, C., & Robinson, K. (2016). Project jacquard: Interactive digital textiles at scale. In *Proceedings of the 2016 CHI conference on human factors in computing systems* (pp. 4216–4227). San Jose: ACM.

Veja, P. (2014). An investigation of integrated woven electronic textiles (e-textiles) via design led processes [PhD thesis, Brunel University, London]. <http://bura.brunel.ac.uk/handle/2438/10528> Accessed 22.04.20.

FIGURE 1
Collection of capacitive and piezoresistive touch and pressure sensor fabrics. © *Pouta, Emmi. (2019) Touch Interwoven. Aalto University, Espoo.*

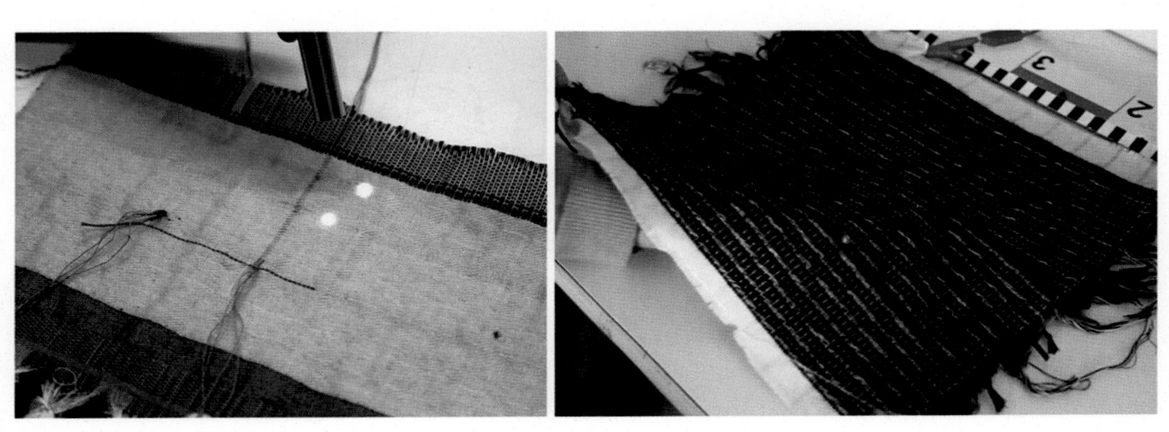

FIGURE 2
Digital textile explorations on magnetic field sensing (left) and microprocessor integration (right). © *Pouta, Emmi. (2016) Woven circuitry. Aalto University, Espoo.*

FIGURE 3

Child interacting with the hand puppet through the fully integrated touch-sensitive user interface. © *Pouta, Emmi. (2019) Interactive hand puppet. Aalto University, Espoo.*

FIGURE 4

The loom provides a stable base for the textile construction versus a free-flowing finished textile. © *Pouta, Emmi. (2019) Touch Interwoven. Aalto University, Espoo.*

Sound as a project requirement: evolution of an experimental tool for psychoacoustic evaluation of materials in architecture and design

Doriana Dal Palù
Department of Architecture and Design, Politecnico di Torino, Torino, Italy

Supervisors: Prof. Claudia De Giorgi and Prof. Arianna Astolfi
Co-supervisors: Dr. Beatrice Lerma and Dr. Eleonora Buiatti

Describing a product sound in simple words, in order to create a clear communication between designers and end-users, is a complex issue (Dal Palù et al., 2018). Each *mechanical* product sound is generated by the interaction between materials, configuration forms, and excitation modes (Dal Palù & De Giorgi, 2018). The relation between sound perception and description and materials experience is, therefore, fundamental. In 2011, an interdisciplinary research group from Politecnico di Torino (Polito)—to which I am part—developed SounBe, an innovative patented tool and method conceived to support designers in their selection of the most suitable materials within the possible hyperchoice, taking into consideration sound as a project requirement. However, an effective validation of the efficacy of SounBe as a tool and method was lacking. My research looked to validate the SounBe tool and method, verify its accuracy as a design tool as well as the effectiveness of the possible research on product sound design to be performed following its method, and to recommend any improvements.

Starting from MATto, the Polito Material Library, I carried out the experimental phase in the anechoic chambers in Polito and IRCAM (Institut de Recherche et Coordination Acoustique/Musique) in Paris. Office chair rolling sounds represented the main case study of the experimentation because of the known relationship between workspace soundscapes and workers' health, job satisfaction, and well-being. More than 90 participants carried out listening tests performed under ecologic, laboratory, and SounBe conditions, comparing the results of the descriptive processes derived from sensorial analysis matter and adopted to give a qualitative characterization to the mechanical sounds. Furthermore, subjective data were compared to objective acoustic and psychoacoustic measurements (i.e., loudness, sharpness, roughness, fluctuation strength, and tonality).

The comparability of sounds obtained with the SounBe method and from the real object in action, which was assumed as a reference, was proved. Consequently, I arranged a set of tests adopting the semantic differential technique, disclosing the possibility to also adopt this technique to sounds obtained with the SounBe tool. The final step of my research involved consultation with enterprises to verify their interest in investing in the technology.

Dal Palù, D., De Giorgi, C., Lerma, B., & Buiatti, E. (2018). *Frontiers of sound in design: A guide for the development of product identity through sounds*. Cham: Springer International Publishing AG.

Dal Palù, D., & De Giorgi, C. (2018). Sound in design: A new disciplinary sub-field? / Il suono nel design: Un nuovo sottoambito disciplinare?, *DIID Disegno Industriale Industrial Design*, *65*, 70−77.

Why is the research needed?

The SounBe method and tool is informed by the sensory analysis techniques adopted to describe wines

Materials Experience 2. DOI: https://doi.org/10.1016/B978-0-12-819244-3.00031-4

and foods, involving the human being as a central element of the evaluation process. It can be applied to every sort of material to provide a new layer of information focused on designing a product sound.

Which aspects of "materials experience" does the research valorize?

This research has contributed to a widening of the materials experience sphere into the domain of "designed sound," that is, intentional sounds from materials that are planned before being produced and reproduced.

How will the research impact on designers?

Sound design, tied to materials sound experience, can be extended from the luxury products domain—traditionally interested in the global quality perception—to the professional and general consumer products domain.

What outcomes have been achieved or are foreseen?

The validation of SounBe, a previously patented tool and method for the psychoacoustic evaluation of materials, allows designers to surf on an increasing set of data on the perceived sound quality of materials, exploring a database of sounds connected with materials, shapes, and gestures—on the one hand—and descriptive and shared labels—on the other hand.

What is the next big challenge for the research area?

After the PhD, a new working prototype of SounBe was developed, with one firm now improving its product sound through the use of the tool and method—a challenge is to increase the commercial uptake.

Which publications most inspired or informed the research?

Byron, E. (2012, October 24). The search for sweet sounds that sell. Household products' clicks and hums are no accident; light piano music when the dishwasher is done? *The Wall Street Journal.* <http://homepages.cae.wisc.edu/~me349/info/sweet_sounds.pdf> Accessed 07.05.20.

Ferreri, M., & Scarzella, P. (2009). *Sound objects, the invisible dimension of design. / Oggetti sonori, la dimensione invisibile del design*. Milan: Electa.

Norman, D. A. (2004). *Emotional design: Why we love (or hate) everyday things*. New York: Basic Books.

Schafer, R. M. (1977). *The soundscape: Our sonic environment and the tuning of the world.* Rochester: Destiny Books.

FIGURE 1

Version of the industrialized SounBe tool. © *Doriana Dal Palù and Giulia Pino.*

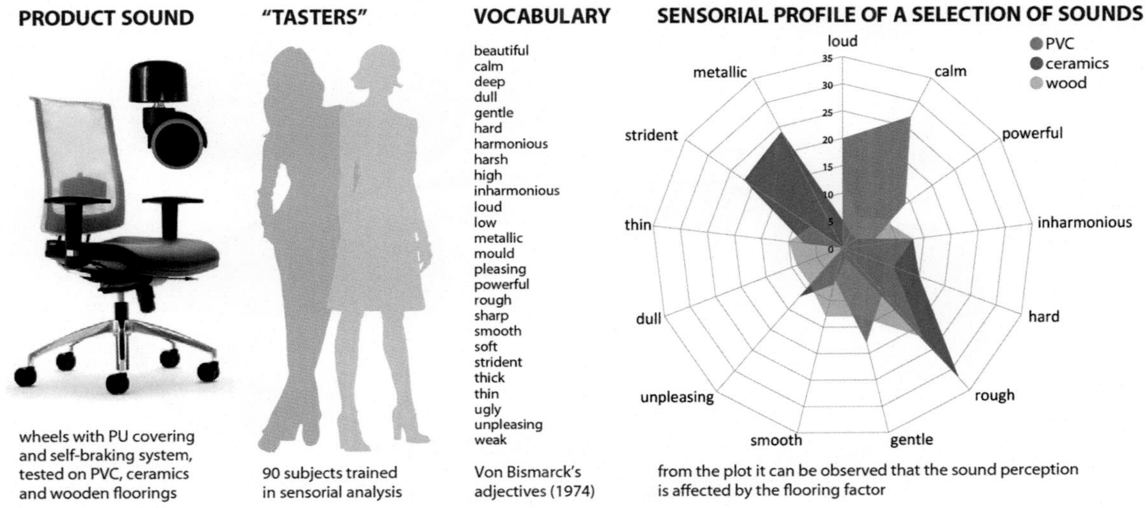

PRODUCT SOUND

wheels with PU covering and self-braking system, tested on PVC, ceramics and wooden floorings

"TASTERS"

90 subjects trained in sensorial analysis

VOCABULARY

beautiful
calm
deep
dull
gentle
hard
harmonious
harsh
high
inharmonious
loud
low
metallic
mould
pleasing
powerful
rough
sharp
smooth
soft
strident
thick
thin
ugly
unpleasing
weak

Von Bismarck's adjectives (1974)

SENSORIAL PROFILE OF A SELECTION OF SOUNDS

PVC
ceramics
wood

loud
calm
powerful
inharmonious
hard
rough
gentle
smooth
unpleasing
dull
thin
strident
metallic

0
5
10
15
20
25
30
35

from the plot it can be observed that the sound perception is affected by the flooring factor

FIGURE 2

A synthetic view of the possible results of a sound design investigation. © *Author.*

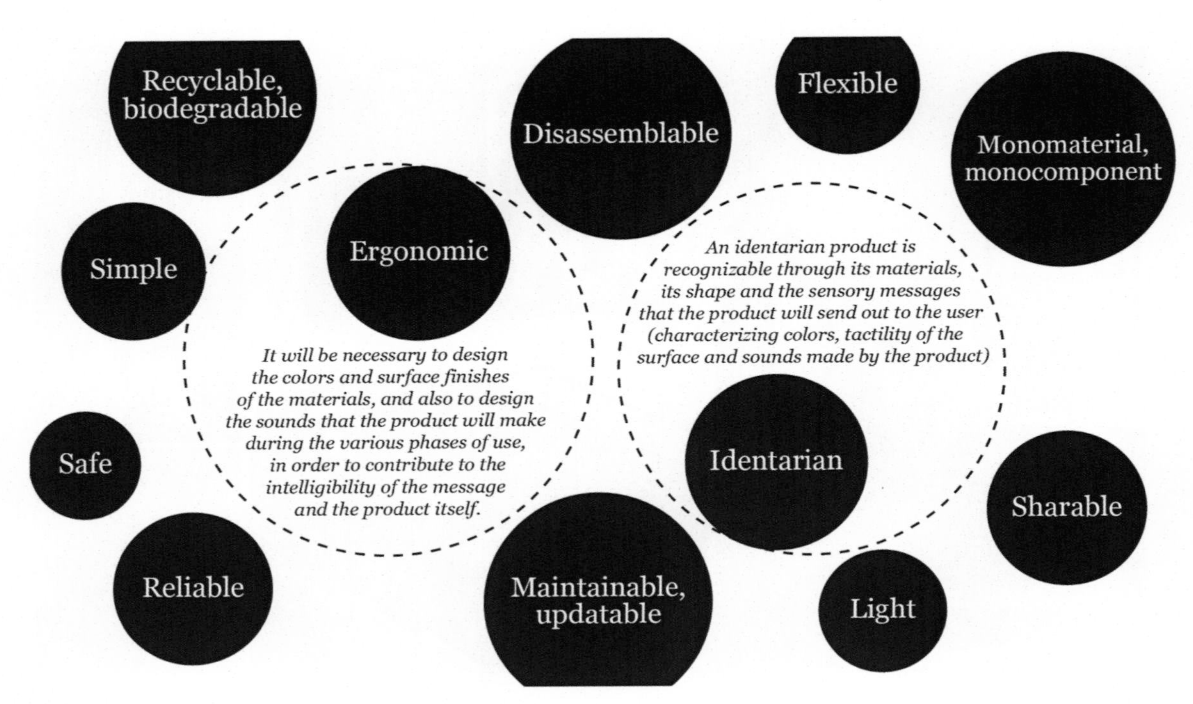

FIGURE 3

Improved set of product requirements, following the sound design integration. © *Doriana Dal Palù and Claudia De Giorgi.*

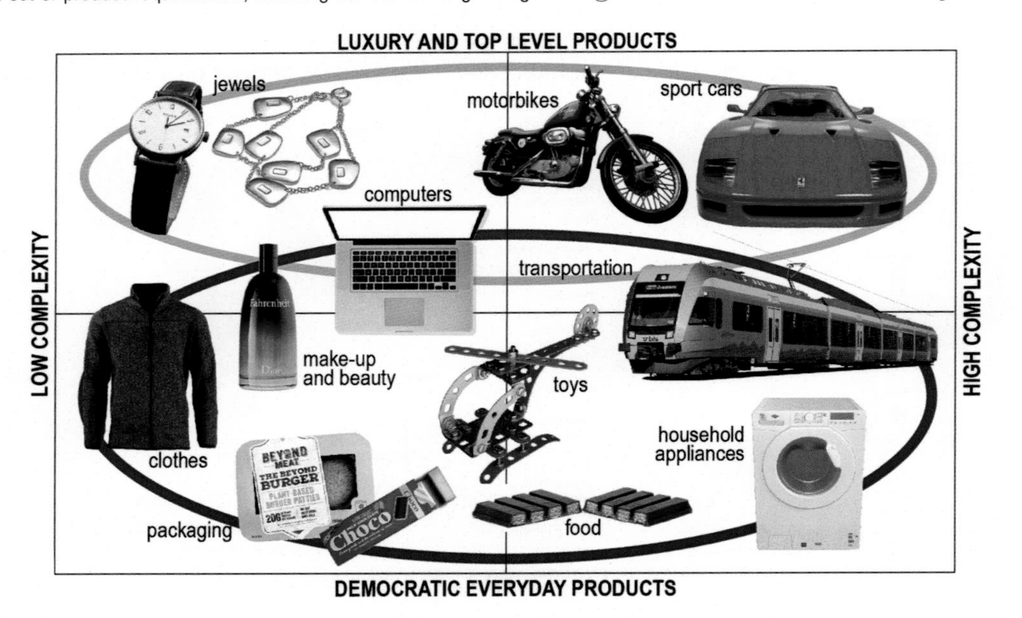

FIGURE 4

The possibility to design product sounds from luxury to everyday products. © *Author.*

Animated puppet skin design: material narratives in visually experienced objects

Vincenzo Maselli
Department of Planning, Design and Technology of Architecture, Sapienza University of Rome, Rome, Italy

Supervisors: Prof. Carlo Martino and Prof. Cecilia Cecchini

In my work, I conducted qualitative research into both puppet animation and the product design field following the three main phases. In the first phase, I outlined a theoretical framework. I brought together research about invisible significances of materials developed in the material design field, in addition to research that values material features in the analysis of animated puppet films. In the former research, I was informed and inspired by the studies of Eleonora Fiorani, Valentina Rognoli, Marinella Levi, Elvin Karana, and others. In the latter, the evocativeness of the surface material qualities of puppets in the analysis of stop-motion films drew upon the work of Suzanne Buchan, Peter Hames, Barry Purves, and Jane Batkin. Unlike the experience of material objects, in the filmic experience users are spectators who interact visually with materials. They experience only the superficial layer of the object, leading me to adopt the ideas of "haptic approach" and "tactile looking" from the field of film studies developed, respectively, by Laura Marks and Jennifer Barker.

In the second phase, I selected short films that belong to a specific subgenre, defined "self-reflexive." In these narratives, the interaction between puppet-makers and materials—that takes place before the shooting—is a relevant element of the diegesis. I analyzed these films by treating puppets as design objects, while the materials and fabrication processes used in the making of the puppets were investigated from a technical point of view. In this phase, explicit and implicit meanings of each film's diegesis were also defined.

In the third phase, an interpretation of symptomatic meanings was made. I established four themes (creation, control, nostalgia, and imperfection) and investigated the aspects through which each short film suggests the themes. In summary, the combined parameters and resources used for the research were as follows: (1) anthropological meanings of used materials, (2) anthropological meanings of manufacturing processes, and (3) relationships between the aforementioned meanings and the diegetic level of the analyzed films.

Maselli, V. (2019). Performance of puppets' skin material: The metadiegetic narrative level of animated puppets' material surface. *The International Journal of Visual Design, 13*(2), 17–37.

Maselli, V. (2018). Puppets animati e materiali: Analisi dei *meta*-racconti di oggetti esperiti visivamente. In R. Riccini (Ed.), *Sul metodo/sui metodi. Esplorazioni per le identità del design* (pp. 327–335). Milano: Mimesis (in Italian).

Why is the research needed?

The research investigates invisible meanings of the material skin of stop-motion puppets, by reframing and connecting theoretical assumptions coming from two fields of studies: product design and film studies. These meanings provide new levels of narrative, and the suggested methodology of analysis ultimately unlocks new interpretations according to a background settled with material, fabrication, and anthropological aspects.

Materials Experience 2. DOI: https://doi.org/10.1016/B978-0-12-819244-3.00002-8

Which aspects of "materials experience" does the research valorize?

The research valorizes the analyses recently developed by design scholars who have studied materials as vehicles of emotions and have focused on concepts such as significances, interpretation, and subjectivity.

How will the research impact on designers?

The work provides a deeper understanding for those involved in puppet design on the role of surface materiality in the manufacturing process of puppets and frames room for improvement applicable to this process.

What outcomes have been achieved or are foreseen?

By merging the cultural references validated by the field of product design in the analysis of fabrication processes and material qualities, with the principles of indirect experience (tactile looking, haptic approach) coming from film studies, the research presents a new set of tools with which to interpret the material aspects of puppets.

What is the next big challenge for the research area?

The four symptomatic meanings associated with the narratives of the material (creation, control, nostalgia, and imperfection) deserve a more thorough study achieve unprecedented results.

Which publications most inspired or informed the research?

Levi, M., & Rognoli, V. (2011). *Il senso dei materiali per il design*. Milano: FrancoAngeli. (in Italian).

Barker, J. (2009). *The tactile eye: Touch and the cinematic experience*. Los Angeles: University of California Press.

Fiorani, E. (2000). *Leggere i materiali: Con l'antropologia, con la semiotica*. Milano: Lupetti. (in Italian).

Marks, L., & Polan, D. (2000). *The skin of the film: Intercultural cinema, embodiment, and the senses*. Durham: Duke University Press.

FIGURE 1

Still frame, *Ab Ovo* (2012) by Anita Kwiatkowska-Naqvi. © *Anita Kwiatkowska-Naqvi.*

FIGURE 2
Still frame, *Ossa* (2015) by Dario Imbrogno. © *Dario Imbrogno.*

FIGURE 3
Still frame, *Komaneko's Christmas: The lost gift* (2009) by Tsuneo Goda. © *Dwarf Studio,* © *TYO/dwarf-Komaneko Film Partners, and* © *Amis de Komaneko.*

FIGURE 4
Still frame, *Walter, a dialogue with the imagination* (2010) by Niels Hoebers. © *Niels Hoebers.*

Material visualization and perception in virtual environments

Mutian Niu

School of Film and TV Arts, Xi'an Jiaotong-Liverpool University, Suzhou, P.R. China,
Department of Civil Engineering and Industrial Design, University of Liverpool, Liverpool, United Kingdom

Supervisors: Dr. Cheng-Hung Lo, Dr. Richard Barrett, Prof. Yue Yong and Dr. Bingjian Liu

Material representation has always been an important part of design visualization. The common method for digital visualization is to adjust the bidirectional reflectance distribution function model to simulate material attributes in the rendered scene. The evaluation of product materials often relies on the rendering outputs of two-dimensional (2D) displays; however, the traditional 2D view only provides stereo images, which cannot give users a strong sense of depth. As an approach to simulate a more realistic environment, virtual reality (VR) strengthens people's immersive experience by delivering a vivid impression of depth. However, the scientific understanding of material perception and its application in VR is still limited. This stimulated us to ask whether the material perception in VR is different from that in "traditional" 2D digital representations, as well as the possibility of using VR as a design tool for facilitating users' material evaluations.

In the first study of this research, I explored differences in material perception between VR and 2D digital material representations. I selected metal and plastic as the test materials, and a cube and sphere as the 3D rendered forms. Thirty participants compared the rendered VR and 2D materials and then evaluated the visual performance of the two digital representation types. In my second study, I explored material perception differences between the two digital representation types within an automobile interior design setting. Taking a car seat as a test model, Substance Designer and Unreal Engine 4 applications were used to deliver renderings of leather and fabric car seats. Roughness and specularity were used as adjustment variables. Participants were exposed to visualizations created in VR and on a 2D screen and asked to give a score for each on a Likert scale against the criteria of glossiness, roughness, reflectance, and saliency. Participants were found to more acutely detect changes in material properties under VR, especially for glossiness and saliency. Participants' more acute detection may be attributed to the visual depth in VR, which will be investigated in future research.

Niu, M., & Lo, C. (2020). An investigation of material perception in virtual environments. In T. Ahram (Ed.), *Advances in human factors in wearable technologies and game design (AHFE 2019)/Advances in intelligent systems and computing Vol. 973* (pp. 416–426). Cham: Springer.

Why is the research needed?

The research investigates to what extent the real-world experience of material perception can be delivered in a virtual environment. In the context of VR technology, studies are made on how the depth and

Materials Experience 2. DOI: https://doi.org/10.1016/B978-0-12-819244-3.00006-5

immersion provided by VR may influence the user's perception of visual qualities of materials.

Which aspects of "materials experience" does the research valorize?

By examining the perceptibility of material qualities in digital renderings, the research extends understanding of the relationship between VR technology and material visualization.

How will the research impact on designers?

The research intends to offer designers and design researchers insights into how to integrate VR technology in the product design process, focused on material decisions.

What outcomes have been achieved or are foreseen?

So far, the research has established perceptual differences between VR and 2D (screen-based) material visualization methods, using standard forms and automotive seating as the experiment stimuli. Design guidelines are being created from the experiment results, aiming to support material selection in design processes making use of VR technology, including the development of suggested good practice.

What is the next big challenge for the research area?

In the near future, evaluation of vivid depth (depth perception of the material authenticity) may provide an important additional dimension of information to extend the use of VR toward the most convincing material appearances.

Which publications most inspired or informed the research?

Fleming, R. (2017). Material perception. *Annual Review of Vision Science, 3*(1), 365–388.

Tang, B., Guo, G., & Xia, J. (2017). Method for industry design material test and evaluation based on user visual and tactile experience. *Chinese Journal of Mechanical Engineering, 53*(3), 162–172 (in Chinese).

Papagiannidis, S., See-To, E., & Bourlakis, M. (2014). Virtual test-driving: The impact of simulated products on purchase intention. *Journal of Retailing and Consumer Services, 21*(5), 877–887.

Sharan, L., Liu, C., Rosenholtz, R., & Adelson, E. (2013). Recognizing materials using perceptually inspired features. *International Journal of Computer Vision, 103*(3), 348–371.

Acknowledgment

This research is partially funded with the Research Development Fund provided by Xi'an Jiaotong-Liverpool University (Project Ref. No. RDF 16-22-02).

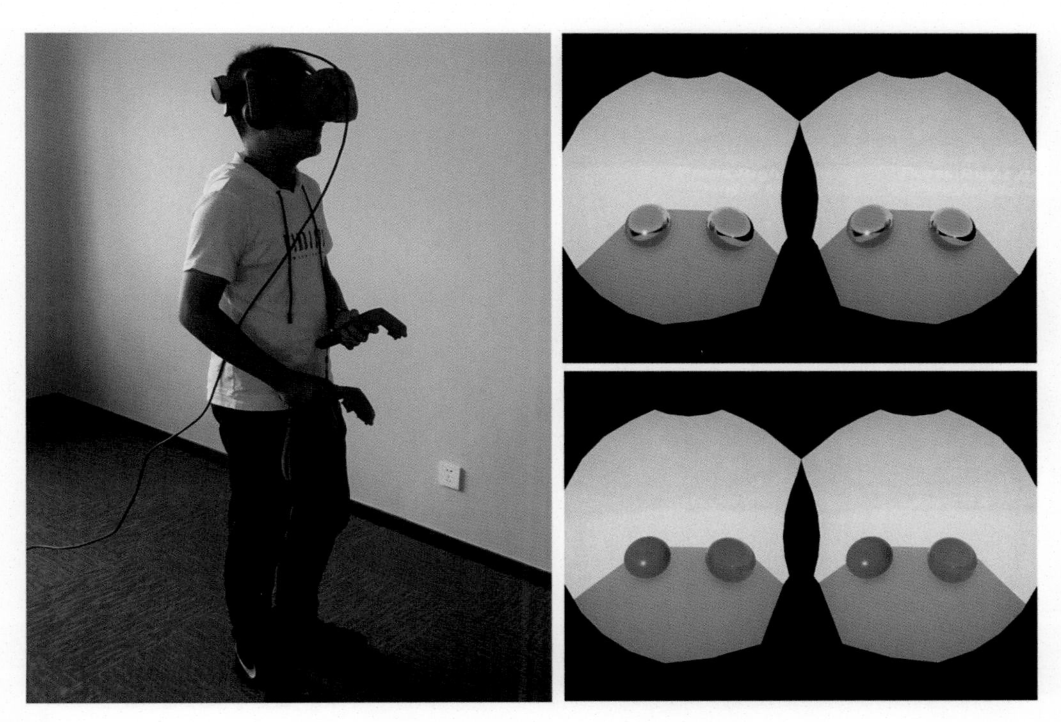

FIGURE 1
Virtual reality (VR) testing process (left) and visual effects in VR (right). © *Author.*

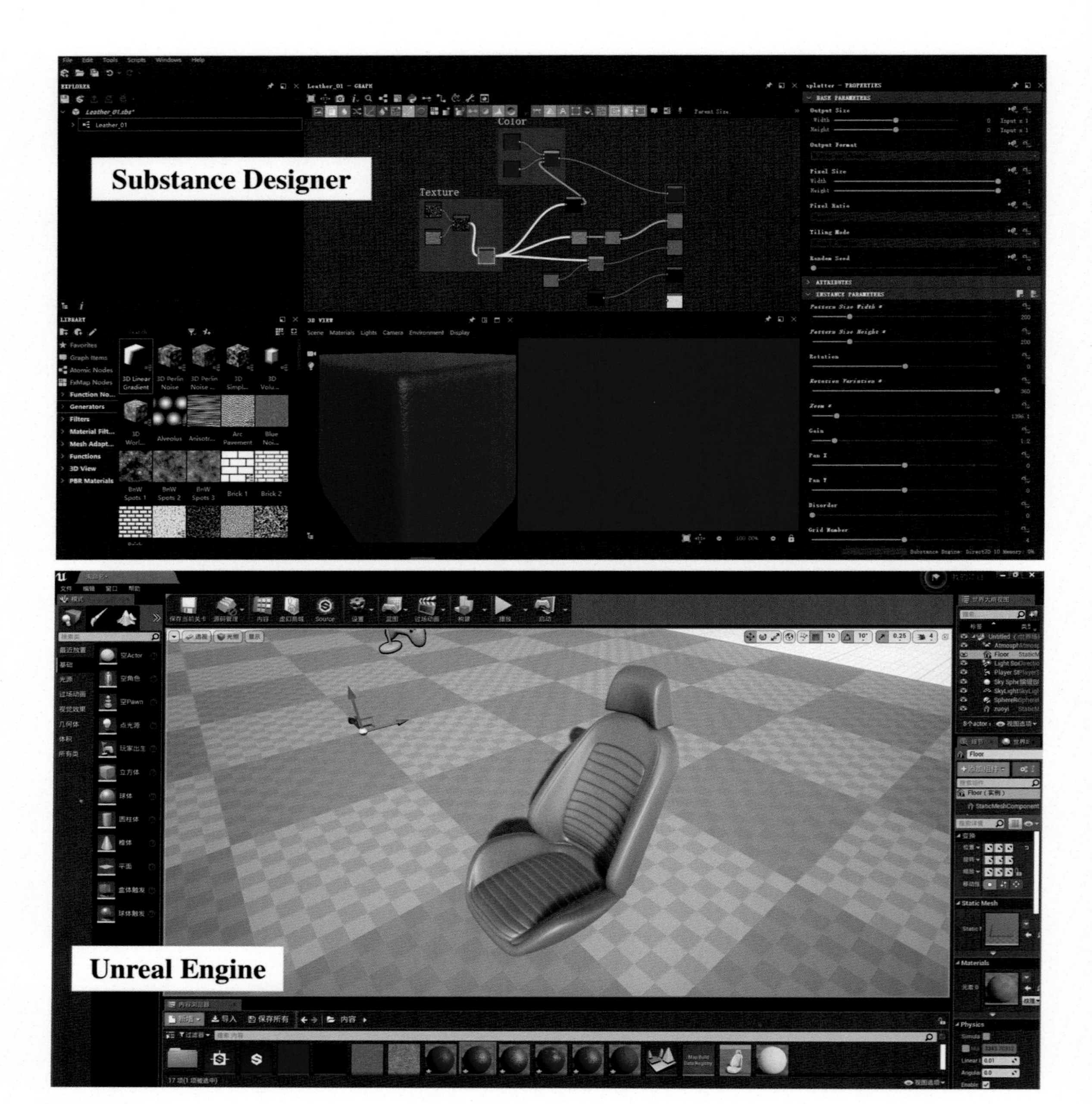

FIGURE 2
Material attribute editing interface and test model in Unreal Engine 4. © *Author.*

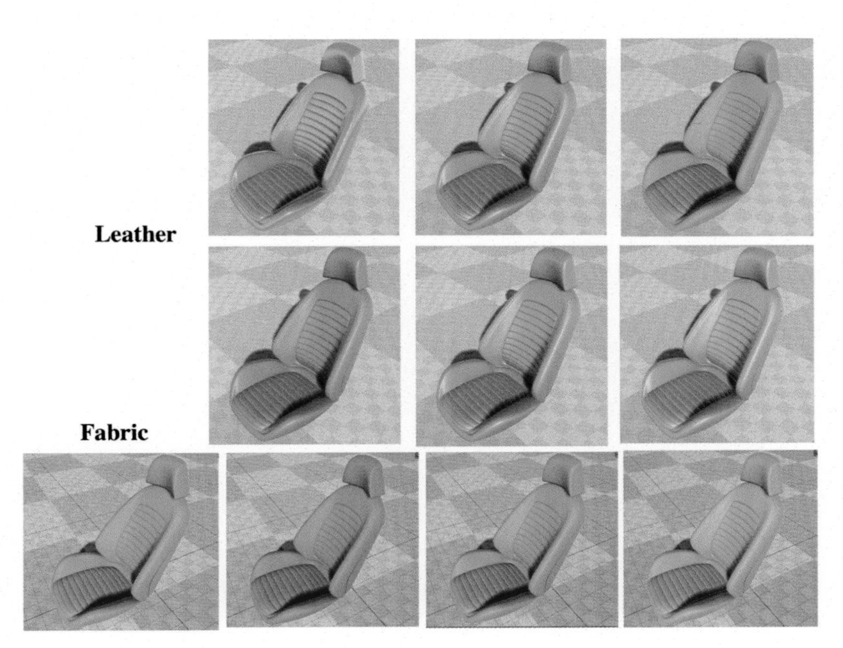

FIGURE 3

Applying materials to 3D models of automobile seats. © *Author.*

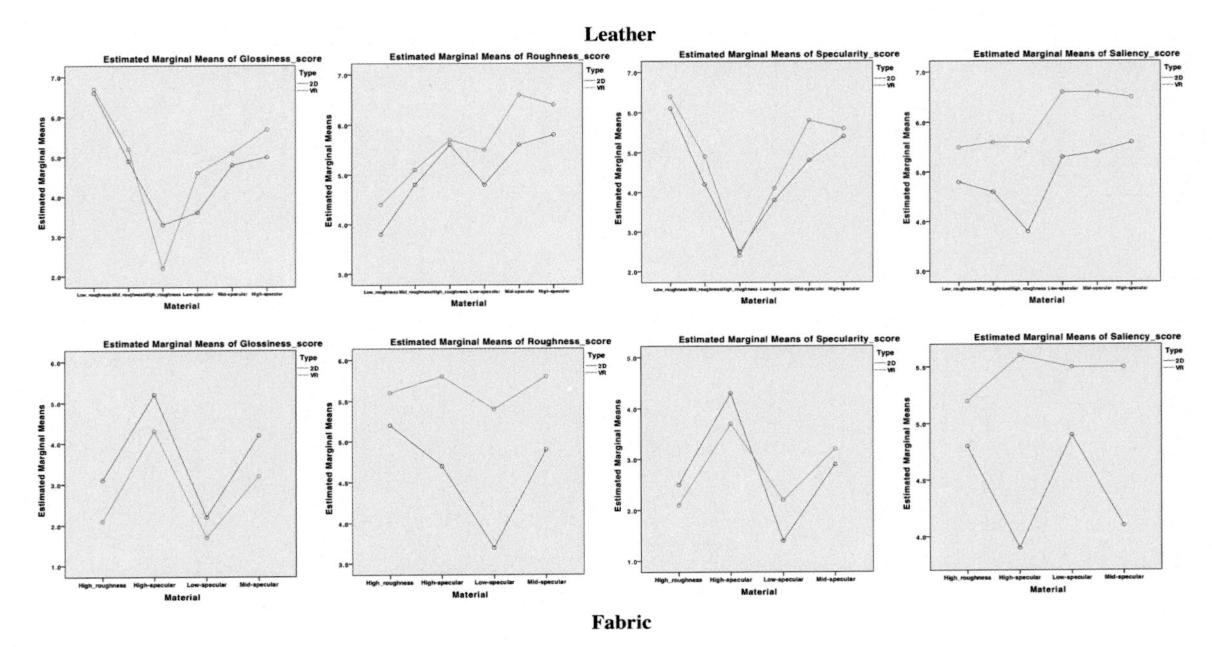

FIGURE 4

Mean plots of main effects on material perception. © *Author.*

End-of-life care through design: visualizing places of death

Michelle Knox
Department of Medicine, University of Alberta, Edmonton, AB, Canada

Supervisor: Assoc. Prof. Aidan Rowe

Death and dying is an increasingly medicalized phenomenon in modern society. Sociologists, designers, architects, and healthcare experts have long been engaged in discussions around patient-centered design in healthcare environments and, by extension, in palliative spaces. While the evidence-based design (EBD) movement has vastly improved hospital and hospice architecture over the decades, relatively little research has applied the EBD principles specifically to buildings associated with terminal illness and death. As such, end-of-life care is too frequently situated within the clinical realm, with its sites of operation often charged with exhibiting the material and nonmaterial features and qualities of medical paternalism.

In this study, I argued that design has the potential and power to disrupt hegemonic medical paradigms and practices—by reframing palliative care as a fundamentally human activity and the palliative environment as a deeply personal realm—where designers might "humanize" what are often seen as primarily "functional" clinical spaces. To do this, I first reviewed relevant scholarly literature, identifying the key gaps between research and practice across architectural, material, and experiential design, particularly noting transformations across locations of death over the decades. Next, drawing upon anthropological perspectives, I analyzed how time and aesthetics manifest in architectural and material artifacts, relate to human anxieties around mortality, and affect individual conceptions and experiences of a "good death." Through ethnographic exercises conducted at a small, independent, suburban hospice and an intensive palliative care unit in an urban hospital, I generated a critical commentary of two very diverse palliative locations. Analyzing research findings from site visits, field observations, photographic material, and interview data, I presented recommendations for the design of future palliative facilities. Finally, findings were contextualized within rapidly changing health landscapes, where future technological, industrial, and social transitions will need further design-led interrogations to continually reconfigure healthcare experiences.

Knox, M. (2021). Locating death anxieties: End-of-life care and the built environment. *Wellbeing, Space and Society, 2*, 100012. https://doi.org/10.1016/j.wss.2020.100012.

Rowe, A., Knox, M., & Harvey, G. (2020). Re-thinking health through design: Collaborations in research, education and practice. *Design for Health, 4*(3), 327–344. https://doi.org/10.1080/24735132.2020.1841918.

Why is the research needed?

This study investigated the effects of built environments on the experience of dying-in-place, with the aim to understand how we might design better material spaces for the end-of-life.

Materials Experience 2. DOI: https://doi.org/10.1016/B978-0-12-819244-3.00004-1

Which aspects of "materials experience" does the research valorize?

This work extends our understanding of how the structural, ambient, and symbolic qualities of materials and spatiality enrich tacit relationships with the places we occupy, particularly facilitating the experience of dying well.

How will the research impact on designers?

The role of designers is often overlooked in healthcare contexts—this work demonstrates how design furthers the objectives of whole-person care and human-centered experience.

What outcomes have been achieved or are foreseen?

A photographic resource has been created, by compiling visuals from field sites and comparing design features of medicalized locations of death. Furthermore, a framework outlining six in-depth recommendations for designing future palliative care sites has been generated, addressing tangible ways in which design should attend to individualized needs for social, cultural, emotional, and sensorial experience.

What is the next big challenge for the research area?

The places where people have been spending their last days have shifted constantly to reflect social, medical, and industrial trends—interdisciplinary scholarship can help designers to better understand the future needs of palliative architecture.

Which publications most inspired or informed the research?

Adams, A. (2016). Home and/or hospital: The architectures of end-of-life care. *Change Over Time, 6*(2), 248–263.

Pallasmaa, J. (2016). Inhabiting time. *Architectural Design, 86*(1), 50–59.

Jencks, C., & Heathcote, E. (Eds.). (2010). *The architecture of hope: Maggie's cancer caring centres*. London: Frances Lincoln Ltd.

Worpole, K. (2009). *Modern hospice design: The architecture of palliative care*. London: Routledge.

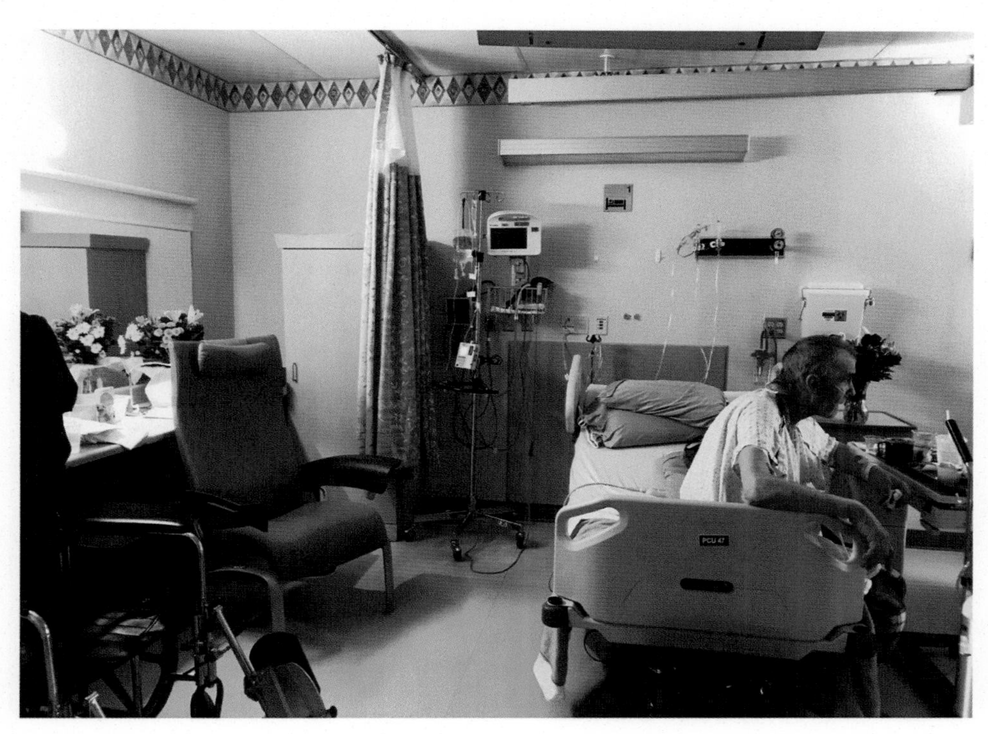

FIGURE 1
Shared room, intensive palliative care unit (general hospital). © *Author.*

FIGURE 2
Communal dining area (hospice facility). © *Author.*

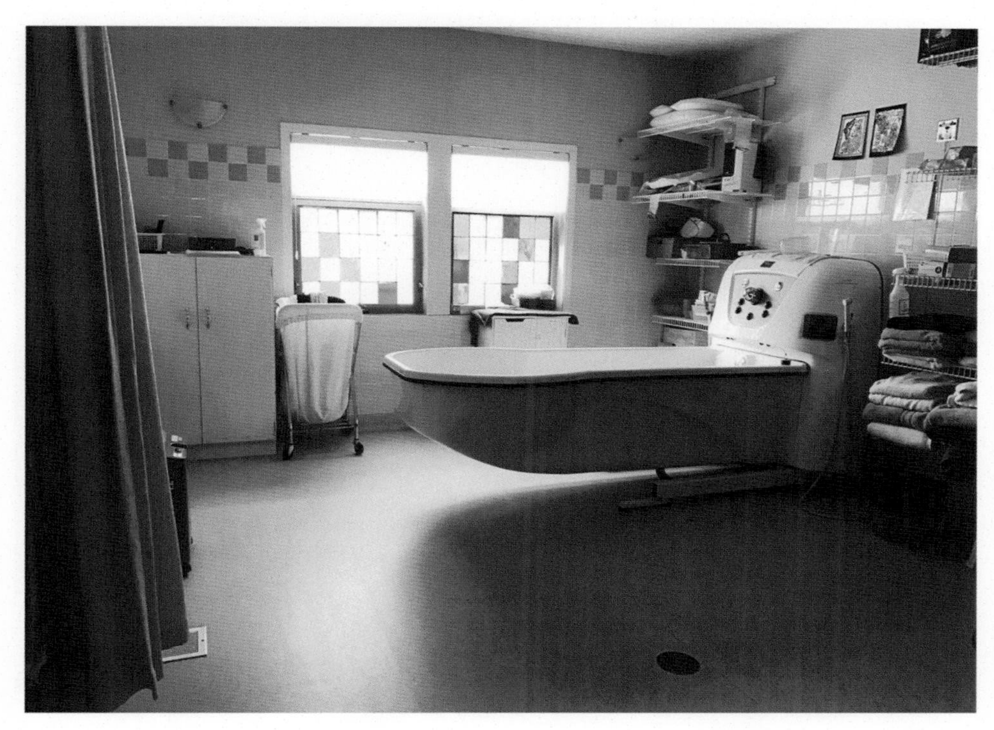

FIGURE 3
Assisted bathing area (hospice facility). © *Author.*

FIGURE 4
Lounge area (hospice facility). © *Author.*

PROJECT 9

Material experiences of menstruation through symbiotic technologies

Marie Louise Juul Søndergaard and Ozgun Kilic Afsar
Media Technology and Interaction Design, KTH Royal Institute of Technology, Stockholm, Sweden

Supervisor: Assoc. Prof. Madeline Balaam

Our research explores how the materiality of menstruating bodies can be used to create new material experiences that stimulate awareness, appreciation, and knowledge about menstrual cycles (Søndergaard et al., 2020). In our collaboration, we bring together knowledge in women's health (Søndergaard, 2018), material science, and design research (Giaccardi & Karana, 2015) to design symbiotic technologies that reconfigure the perception of bodily fluids as body waste toward a material that can be appreciated and repurposed. We experimented with repurposing menstrual blood using various technologies and DIY (do it yourself) scientific instruments to facilitate a deeper engagement for the menstruator. We prototyped three encounters with menstrual blood (tactile, acoustic, and visual).

The tactile encounter involved soft robotic actuators. We designed soft on-skin actuators made from silicone with a variety of shapes and patterns actuated by air or blood. The initial prototypes are pneumatic units with preprogrammed air movement that react to pressure exerted on the material surface, rendering a predefined library of haptic feedback. Similar forms are utilized with hydraulic actuation. The purpose of these sensor-actuator couplings is to bring attention and sensitize the body to the felt experiences of menstruation whether it is a cramp, a throbbing pain, or else a feeling of flow and release. The acoustic encounter involved Helmholtz resonators and blood. By exploiting the acoustic properties of the menstrual cup, we explored sonic encounters with collected menstrual blood. First, we envisioned a sonic performance through the changing resonant frequency of the menstrual cups as the blood is collected. Second, we measured and monitored the volume of blood that is lost during a single cycle by reverse-calculating the volume from the resonant frequency of the vessel. Finally, the visual encounter utilized wearable microfluidics. We speculated on a design scenario using microfluidics in which a wearable chip provides the facility of an informative color-tracking tool throughout the period. This application explores how a more accessible control of the biofluid could support menstruators in understanding and repurposing their own menstrual blood. As an initial step, we probed a DIY microfluidic device to unleash this low-cost instrument's interactive potential for designers beyond medical diagnosis.

Woytuk, N., Søndergaard, M., Felice, M., & Balaam, M. (2020). Touching and being in touch with the menstruating body. In *Proceedings of the 2020 CHI conference on human factors in computing systems (CHI'20)* (pp. 1–14). New York: ACM. https://doi.org/10.1145/3313831.3376471.

Søndergaard, M., Afsar, O., Felice, M., Woytuk, N., & Balaam, M. (2020). Designing with intimate materials and movements: Making "menarche bits." In *Proceedings of the 2020 DIS conference on designing interactive systems (DIS '20)*. New York: ACM. https://doi.org/10.1145/3357236.3395592.

Materials Experience 2. DOI: https://doi.org/10.1016/B978-0-12-819244-3.00007-7

Why is the research needed?

The research explores how menstrual blood and menstruating bodies can become materials in designing symbiotic technologies that make space for experiences of menstruation. The purpose is to design technologies that enable a menstruator to engage with their menstrual blood in ways that move it beyond a waste product.

Which aspects of "materials experience" does the research valorize?

The research valorizes menstrual blood as something to touch and explore (Campo Woytuk et al., 2020); a material full of interactional and energetic opportunities that have yet to be fully explored.

How will the research impact on designers?

Designers of women's health technologies are intended to be impacted by the research, working closer to/with the materials of the menstruating body to create and deepen situated knowledge of the body (Höök, 2018).

What outcomes have been achieved or are foreseen?

Three artifacts have been designed that reconfigure the material experiences of menstruation. The artifacts change the perception of menstrual blood from being a waste product to a design material that allows people to touch, hear, and see it in ways that make them feel comfortable, curious, and confident in their body.

What is the next big challenge for the research area?

The next big challenge is to move beyond proof of concept technologies to robust and reliable designs, which can be used sustainably over the longer term.

Which publications most inspired or informed the research?

Höök, K. (2018). Designing with the body: Somaesthetic interaction design. In K. Friedman & E. Stolterman (Eds.), *Design thinking, design theory* (p. 272). Cambridge: MIT Press.

Giaccardi, E., & Karana, E. (2015). Foundations of materials experience: An approach for HCI. In *Proceedings of the 33rd annual ACM conference on human factors in computing systems (CHI'15)* (pp. 2447–2456). Seoul: ACM Press. https://doi.org/10.1145/2702123.2702337.

Søndergaard, M. (2018). *Staying with the trouble through design: Critical-feminist design of intimate technology* [PhD thesis, Aarhus University, Aarhus]. https://doi.org/10.7146/aul.289.203.

Acknowledgments

This research was funded by the Swedish Research Council under Project 2017-05133.

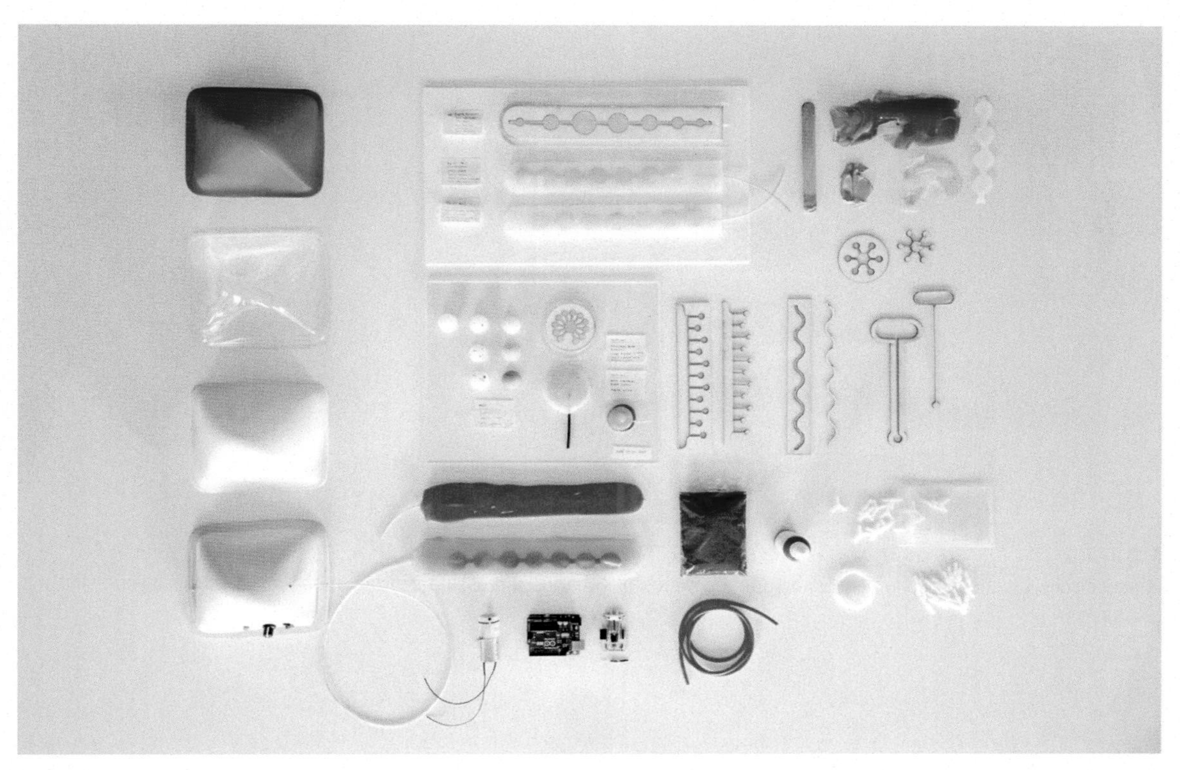

FIGURE 1

Materials involved in the soft robotic actuators: silicone, electronics, tubing, fittings, casing, molds, and stencils. © *Authors.*

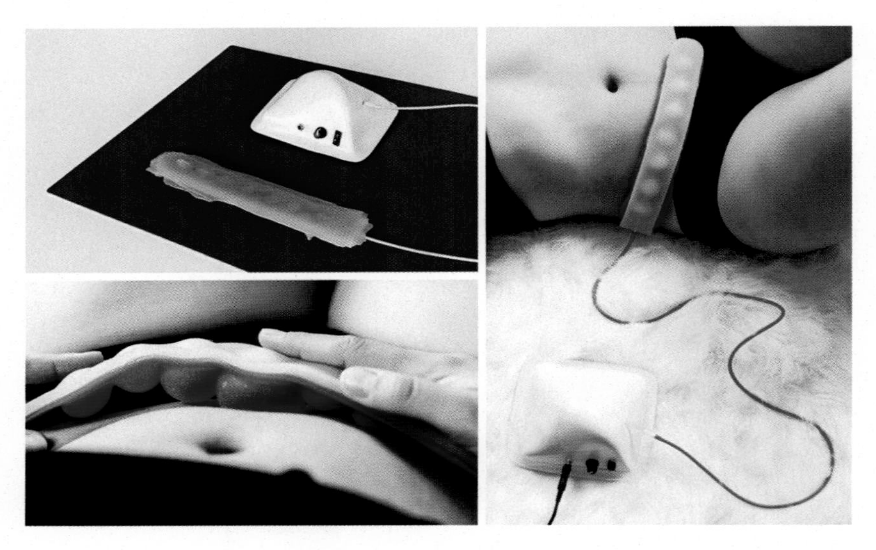

FIGURE 2
Pneumatically actuated silicone parts with battery-powered unit that uses pressure-sensing to control air flow inflation/deflation.

FIGURE 3
Scientifically blown glass Helmholtz resonators preserving the menstrual blood throughout one menstrual cycle. © *Authors.*

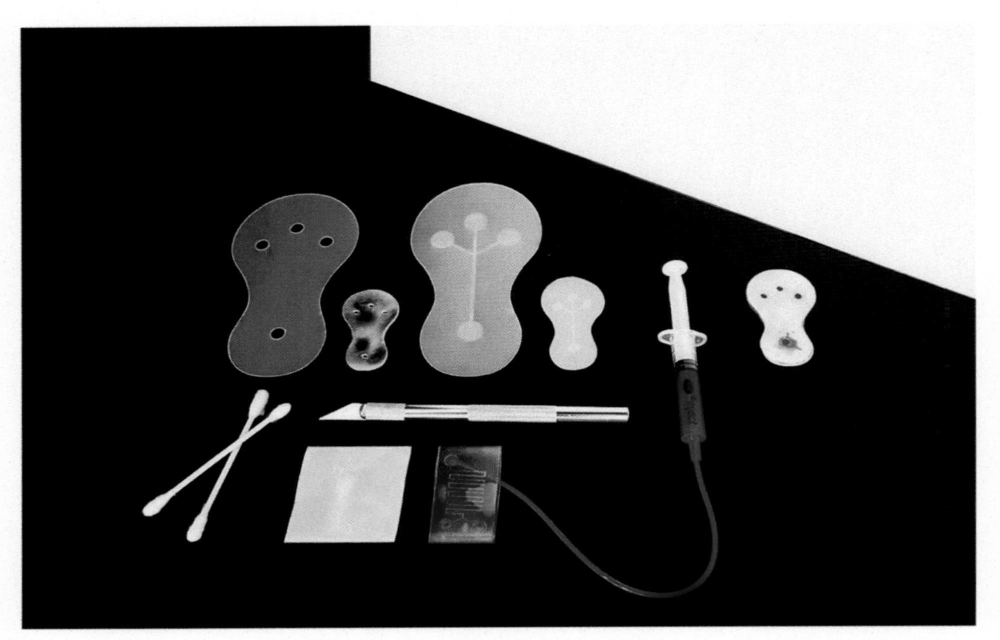

FIGURE 4
DIY microfluidics speculating on the analysis of chemical and biological elements of menstrual blood. © *Authors.*

PROJECT 10

The salt material house project: designing for death

SunMin May Hwang
Human Factors and Ergonomics, University of Minnesota, Minneapolis, MN, United States of America

Supervisor: Brandon Clifford

In the conventional practice of architecture, "obsolescence over time" is undesirable. In this project, obsolescence over time is one of the essential elements, centered on construction and demolition. I carried out the project by designing a salt-cured seasonal residence, which will gradually and naturally be demolished over a designated period through natural weathering agencies such as wind, water, and degradation. The building life expectancy will be precisely set out from the beginning to the end, purporting each step of its life cycle, from occupation to demolition. Some parts of the building will obviously remain for a longer period, depending on its structural integrity.

As a result, the big picture is that the residence will evolve over time, varying not only in its form but also in its function. The material deformation will provide subtle changes at different times of the day and year, which will then cause to serve different functions at different stages of the building's lifespan. To implement this idea, I used building materials all from natural resources including salt, soil, gravel, sand, and coconut fiber. Water and heat are the binding solution of the building structure, whereas wind and rain serve as the demolition agents.

Regarding the building form, reposed mounds built out of earth will define the formwork (or mold) for the house. Using the angle of repose as the formwork allows the internal space to follow the size of the mound. Additional material—coconut fiber—is layered on top of each mound to become the building façade. Salt crystallization over this layer will add rigidity and controlled opacity to the house. The important aspect of this method is that it leaves no harmful impact on the ecology. Forms will vary in diameters, heights, and angles depending on the mixture of earthen elements, lending uniqueness and variability each time a house is built with this method.

Why is the research needed?

We as architects consider ourselves creators. Yet, we work under the false assumption that buildings will last forever. This project challenges the conventional notion of building sustainability. The research illustrates a proof of concept for a redefined meaning of sustainability by reversely mandating a building's life expectancy.

Which aspects of "materials experience" does the research valorize?

The reversal of the concept of sustainability gives rise to a whole new palette of materials, such as natural agents, without compromising the avoidance of adverse impacts on the environment.

Materials Experience 2. DOI: https://doi.org/10.1016/B978-0-12-819244-3.00023-5

How will the research impact on designers?

Not only should this project invoke persistent effort in testing new materials, but more importantly, it should also stimulate a shift toward new paradigms of sustainability in our design practices.

What outcomes have been achieved or are foreseen?

The research has evoked a quintessentially unique materials experience to the convention by shifting the way we, as architects, operate. Newly defining building sustainability, designing for death as opposed to perpetuity, has sparked a vitality to achieve eccentric expressions of lifestyles through the unique selection of natural materials and the use of natural forces for construction and deconstruction.

What is the next big challenge for the research area?

Challenges lay in the scaling-up from proof of concept, specifically on the impediments of shifting to a new paradigm of sustainability—more so than the replication of the method.

Which publications most inspired or informed the research?

Ness, D. A. (2019). *The impact of overbuilding on people and the planet.* Newcastle-Upon-Tyne: Cambridge Scholars Publishing.

Guy, B., & Ciarimboli, N. (2008). *DfD: Design for disassembly in the built environment: A guide to closed-loop design and building*. Hamer Center for Community Design, Pennsylvania State University.

Crowther, P. (2003). *Deconstruction: Techniques, economics, and safety*. Brisbane: Queensland University of Technology.

Kibert, C. J., Chini, A. R., & Languell, J. (2001). Deconstruction as an essential component of sustainable construction. In J. Duncan (Ed.), *Proceedings of CIB world building congress* (Paper: Nov 54). Wellington: Branz.

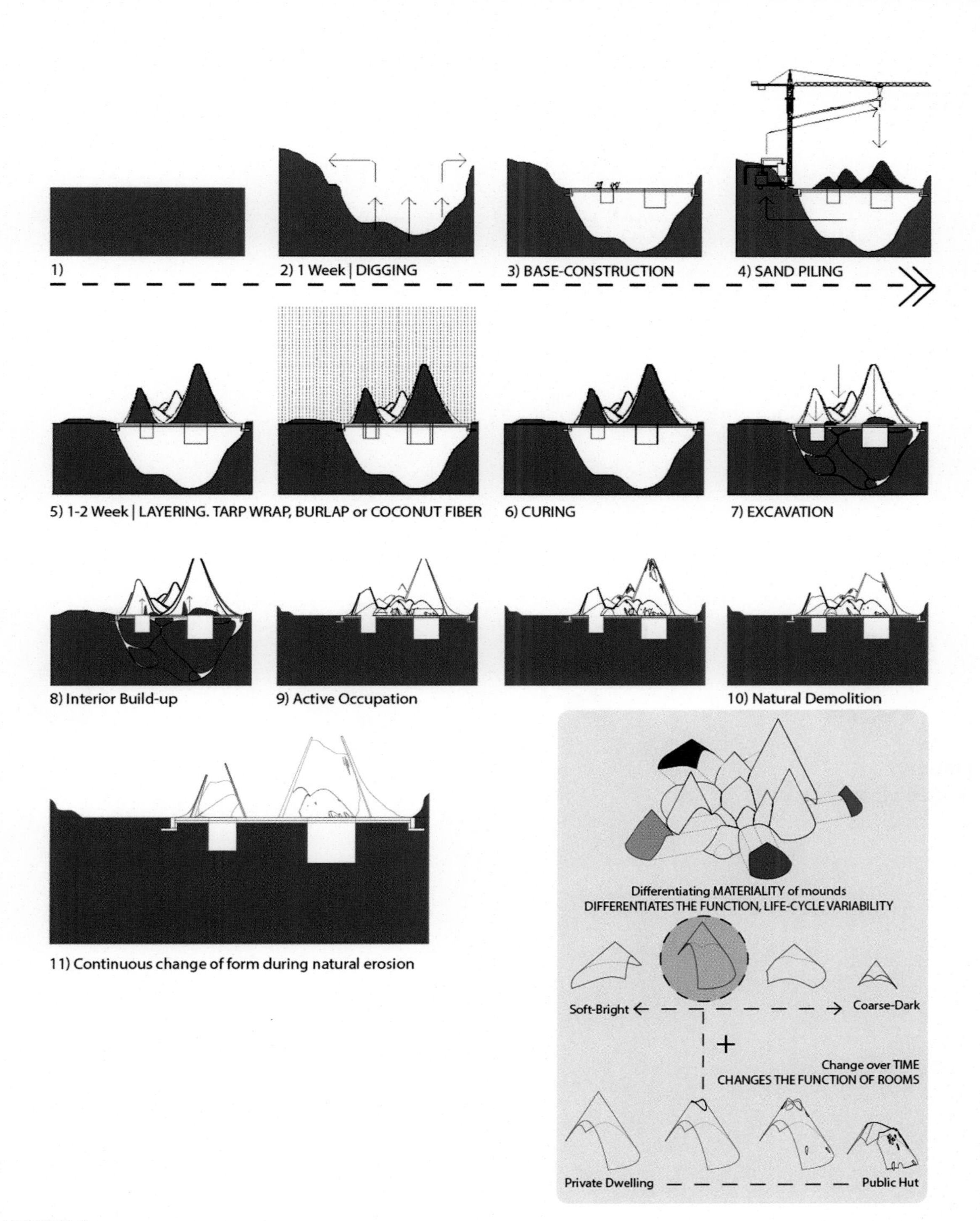
1)
2) 1 Week | DIGGING
3) BASE-CONSTRUCTION
4) SAND PILING
5) 1-2 Week | LAYERING. TARP WRAP, BURLAP or COCONUT FIBER
6) CURING
7) EXCAVATION
8) Interior Build-up
9) Active Occupation
10) Natural Demolition
11) Continuous change of form during natural erosion
Differentiating MATERIALITY of mounds
DIFFERENTIATES THE FUNCTION, LIFE-CYCLE VARIABILITY
Soft-Bright
Coarse-Dark
Change over TIME
CHANGES THE FUNCTION OF ROOMS
Private Dwelling
Public Hut

FIGURE 1
Building construction diagram. © *Author.*

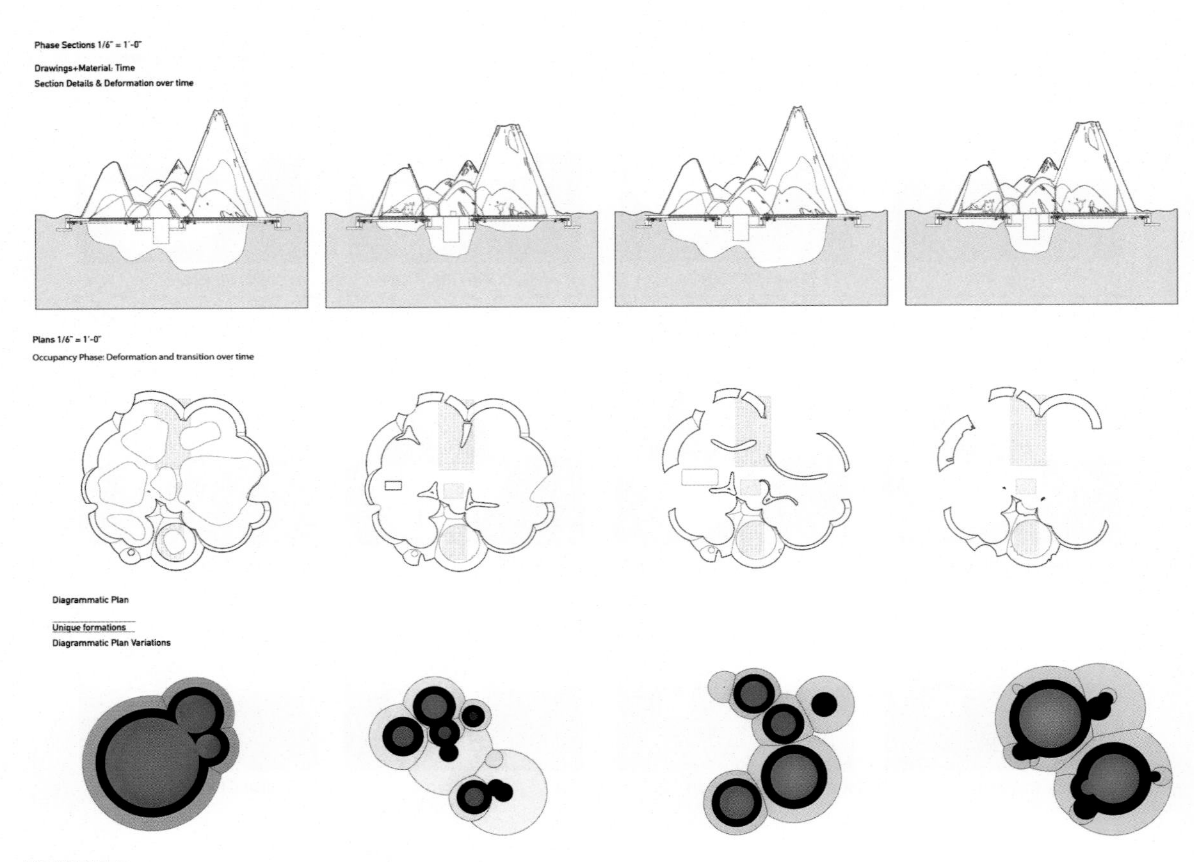

FIGURE 2
Life cycle architectural drawings. © *Author.*

MATERIAL TEST: Salt Crystallization
Coconut Fiber | Basswood | Corrugated Cardboard | Cotton Fabric

SALT SOLIDITY TEST:
Oven Baking | Boiling | Microwave | Natural Evaporation

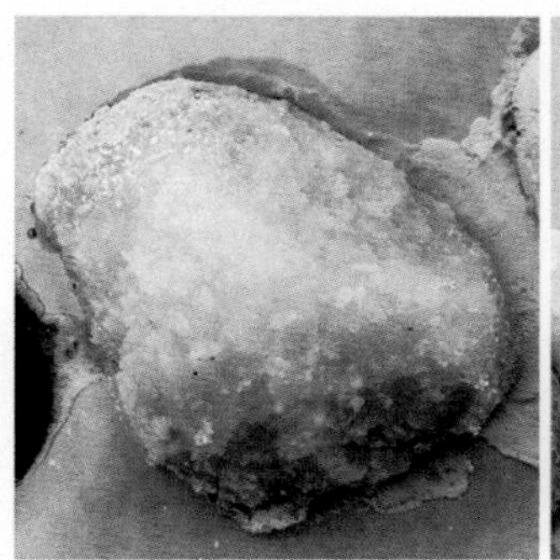
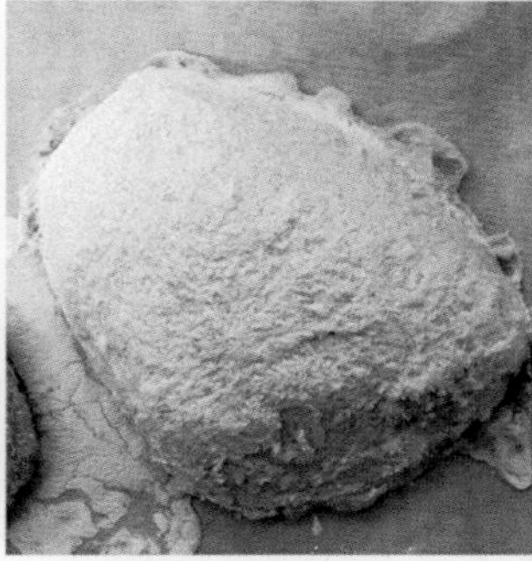

DEMOLITION TEST:

Agent:
Rain water

Building Material:
Cured Salt +
Coconut Fiber

FIGURE 3
Material experiments. © *Author.*

FIGURE 4
Exterior view of The Salt Material House Project. © *Author.*

PROJECT 11

Reflecting on material interactions as a way of being with the world

Bilge Merve Aktaş[a] and Camilla Groth[b]
[a]Department of Design, Aalto University, Espoo, Finland,
[b]Department of Visual and Performing Arts Education, University of South-Eastern Norway, Notodden, Norway

In this research, we present the Human–Material Interaction course that was offered to an interdisciplinary group of undergraduate university students. In this course we used craft-making as a pedagogical platform to be able to reflect on materiality and its extended meanings to us in a concrete and experiential manner. The students, who were novices to a large degree, reflected on material interactions and shared their experiences in the group. The course incorporated both theoretical lectures and hands-on creative studio sessions. In the lectures, the teachers, who are also the authors of this project, presented recent theories in the field of material studies and experiential knowledge. The approach draws on the works of Tim Ingold, Lambros Malafouris, Jane Bennett, and others in this vein.

In the studio sessions, students worked with clay or wool and familiarized themselves with these materials through experimentation. Students documented their processes by utilizing a reflective diary. By facilitating discussions in class, we examined how using first-hand experiences of materials, gained through craft practice, enabled reflections on everyday material interaction, paying extra attention to our dependency on and coexistence with various materials. Through this study, we discuss how experiential knowledge about materials, gained through craft practice, can become a medium for critically reflecting on the impact of our everyday material interactions and, on a large scale, contemporary environmental issues. The students' reflections indicate that gaining first-hand experiences renews their relationship with materials toward a more dialogical one, in which humans do not dictate and exploit materials but rather follow a more responsible interaction.

Aktaş, B.M., & Mäkelä, M. (2019). Negotiation between the maker and material: Observations on material interactions in felting studio. *International Journal of Design, 13*(2), 55–67.

Groth, C. (2017). *Making sense through hands: Design and craft practice analysed as embodied cognition* [Doctoral dissertation, Aalto University, Aalto ARTS Books, Helsinki].

Why is the research needed?

The research examines human–material interaction through first-hand experiences to facilitate a discussion about how humans collaborate with materials rather than only use them as a means to an end. Investigating the material interaction processes is needed to gain an understanding of coexistence, sustainability, and respect for materiality in general.

Which aspects of "materials experience" does the research valorize?

Materials experience here tackles materials as active entities and refers to experiences with and of raw

Materials Experience 2. DOI: https://doi.org/10.1016/B978-0-12-819244-3.00012-0

materials and their qualities, affordances, and resistances and to the meanings generated during the interaction.

How will the research impact on designers?

The pedagogical method and the theoretical discussion can navigate designers when they reflect on their own relationship with materiality to see their personal role, position, and behavior in material interactions.

What outcomes have been achieved or are foreseen?

The research has resulted in a pedagogical method that utilizes material-based craft-making as a platform to facilitate interdisciplinary discussions on societal challenges and human behavior.

What is the next big challenge for the research area?

In the next phases, the challenge is to find ways to share knowledge gained through materials experiences and craft-making with other disciplines, to explore potential crossovers and benefits.

Which publications most inspired or informed the research?

Mäkelä, M. (2016). Personal exploration: Serendipity and intentionality as altering positions in a creative practice. *FORMakademisk, 9*(1), 1–12.

Ingold, T. (2013). *Making: Anthropology, archaeology, art and architecture*. New York: Routledge.

Malafouris, L. (2013). *How things shape the mind: A theory of material engagement*. London: MIT Press.

Nimkulrat, N. (2012). Hands-on intellect: Integrating craft practice into design research. *International Journal of Design, 6*(3), 1–14.

FIGURE 1
At the beginning of the course, each student worked at the potter's wheel while blindfolded to grasp the material's behavior. © *Authors.*

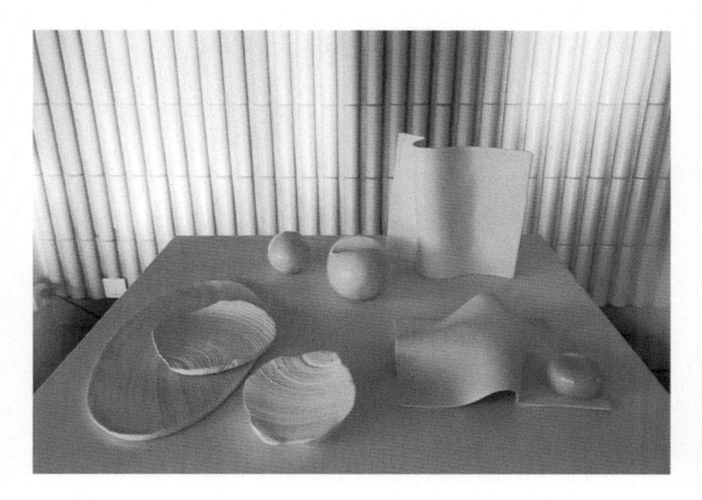

FIGURE 2
A design student worked with clay to develop material-based textures rather than textures dictated by the maker. © *Authors.*

FIGURE 3
A computer science student developed a form based on material features, examining the durability of wool fibers. © *Authors.*

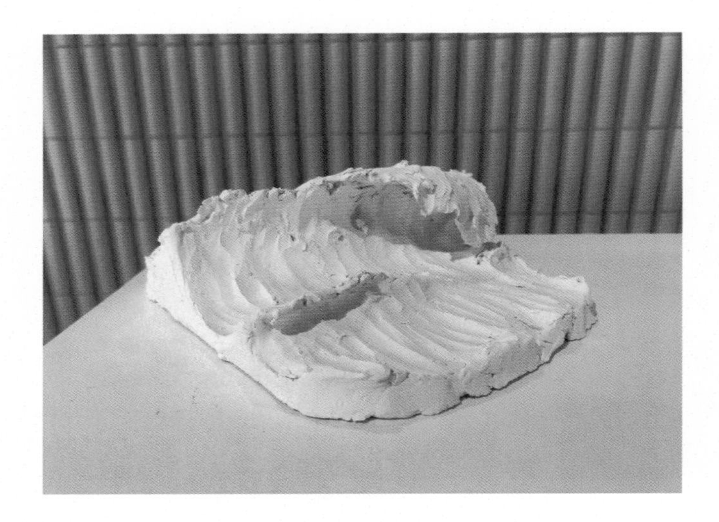

FIGURE 4

A business student worked with the feelings that the clay evoked in her, such as experiences of sandy beaches, and made a form that embodied material connotations from previous experiences with the material. © *Authors.*

PROJECT 12

Beyond biomimicry: developing a living building realm for a postanthropocene era

Assia Stefanova
School of Architecture, Planning and Landscape, Newcastle University, Newcastle upon Tyne, United Kingdom

Supervisors: Prof. Rachel Armstrong and Dr Ben Bridgens

The built environment that we have fashioned for ourselves is predicated on separating humanity from the rest of nature through the use of impermeable manmade barriers. My research examines the potential for a new living technology in the form of living materials that interface with nature and our own ecology, mediating between the two and selectively facilitating exchanges. I use microorganisms to create biocomposite materials that have living metabolic functions for the useful part of their lifecycle. These materials are capable of feeding on waste products from our environment and are able to improve the air quality of building interiors, reducing the need for mechanical ventilation. My research challenges modern ideas of hygiene and proposes ways of creating living modules that can be integrated within our buildings that would allow other species to flourish. The materials developed provide healthier environments for human habitation while encouraging a more inclusive built environment.

My material testing involves the use of cross-disciplinary laboratory practices that follow protocols for assessing the viability of the biocomposites regarding compatibility between substrate and organism, longevity, and the ability to function within an open setup. The material development is accompanied by experimental digital fabrication in the form of 3D printing. This aspect of my work is focused on the development of new ways of fabricating with living materials and the generation of components that can be scaled for mass production.

Through developing living materials, we develop networks of collaboration within research practices, but more importantly we establish collaborative environments between species, where the designer assumes the role of an orchestrator of desirable behavior of living biota, by providing encouragement in the form of conditions for those microorganisms to live and develop in a favorable manner. By working with living biota, we accept a degree of responsibility and establish a practice of care that is vital in this emerging field and that distinguishes it from working with inanimate matter, hence creating a new material practice.

Stefanova, A., In-Na, P., Caldwell, G. S., Bridgens, B., & Armstrong, R. (2020). Architectural laboratory practice for the development of clay and ceramic-based photosynthetic biocomposites. *Technology|Architecture + Design,* 4(2), 200–210. https://doi.org/10.1080/24751448.2020.1804764.

Stefanova, A., In-Na, P., Caldwell, G. S., Bridgens, B., & Armstrong, R. (2021). Photosynthetic textile biocomposites: Using laboratory testing and digital fabrication to develop flexible living building materials. *Science and Engineering of Composite Materials, 28*(1), 223–236. https://doi.org/10.1515/secm-2021-0023.

Why is the research needed?

It is important to develop new testing methods and set out ways of conducting interdisciplinary research that ventures outside of established design practices in order to be able to effectively participate in the development of living materials. By engaging with materials

Materials Experience 2. DOI: https://doi.org/10.1016/B978-0-12-819244-3.00016-8

at such a fundamental level, we are able to introduce an ethical aspect to the work that challenges established perceptions of matter.

Which aspects of "materials experience" does the research valorize?

The work focuses on the development of biological material alternatives and their integration into the building realm and by extension into new material experiences. It explores ideas of cohabitation and codesign through collaborative human/nonhuman interactions.

How will the research impact on designers?

The research will present a methodological blueprint for developing living materials and creating systems for sustaining their metabolic functions.

What outcomes have been achieved or are foreseen?

The aim of the research is to develop a system that eliminates pollutants through the use of natural metabolic processes as well as to foster a relationship between the human and nonhuman species as a framework for sustainable coexistence.

What is the next big challenge for the research area?

Sustaining highly specialized organisms within an uncontrolled setting presents challenges of managing contamination in a sustainable manner as well as cultivating maintenance practices that encourage a feeling of responsibility within inhabitants.

Which publications most inspired or informed the research?

Hird, M. J. (2009). *The origins of sociable life: Evolution after science studies*. New York: Palgrave Macmillan.

Merchant, C. (1980). *The death of nature: Women, ecology, and the scientific revolution*. San Francisco: Harper and Row.

Proksch, G. (2018). Built to grow: Blending architecture and biology. *Technology | Architecture + Design, 2*(1), 118–119.

Armstrong, R. (2015). *Vibrant architecture: Matter as a codesigner of living structures*. Warsaw: De Gruyter Open.

FIGURE 1

Living mycelium on glazed ceramic, by Assia Stefanova and Dilan Ozkan (Yggdrasil Exhibit, London Design Festival, 2019). © *Author.*

FIGURE 2

Ceramic scaffolds for growing algae in repeated cycles by Assia Stefanova. © *Author.*

FIGURE 3
Biocomposite prototype of living algae on porcelain, by Assia Stefanova (Yggdrasil Exhibit, London Design Festival, 2019). © *Author.*

FIGURE 4
Chlorella vulgaris, microalgae growing on a multifaceted ceramic laden with nutrient media, by Assia Stefanova (Yggdrasil Exhibit, London Design Festival, 2019). © *Author.*

PROJECT 13

Healing materialities from a biodesign perspective

Barbara Pollini
Design Department, Politecnico di Milano, Milan, Italy

Supervisor: Assoc. Prof. Dr. Valentina Rognoli

The shift from a linear economy to a circular one still has a long way to go, but innovative and disruptive design approaches are already emerging. In this context, the focus of my research is on "healing materialities," meaning those material scenarios based on the regenerative processes of resources instead of depletion. Including both living materials (made with living organisms) and life-enabling materials (e.g. bioreceptive ones), my research aims to define the boundaries of newly designed materialities where the final goal is to support life.

This research intersects the constantly evolving concept of sustainability, the material design discipline, and biodesign—the latter being a radical approach based on the integration of living organisms as functional components in a design process. In my work I apply a transdisciplinary approach to understand the implications that living materials can have on sustainable design and behavior change, mainly focusing on the relationships among living and nonliving agents, alongside a more-than-human design perspective.

Which materials can be defined as "healing" in a sustainable design context? Which material features become relevant, introducing the variable of living organisms into design? How can designers enhance living/nonliving interactions to support life? To answer these research questions, I am collecting and analyzing qualitative data through a mixed-methods research: reviewing projects, carrying out expert interviews, and experimenting with various biofabricated materials using a biotinkering approach.

Interest toward biodesign has increased substantially in the last 10 years, yet a dedicated methodology and structured studies on the field are still to come. Focusing on biofabricated materials, my research aims to define new approaches and guidelines to design life-enabling materialities, grounded on material-driven design, sustainable design (including life-cycle design and circular design), biodesign, biomimicry, multispecies and more-than-human design, and speculative design.

Pollini, B. (2020). Sustainable design, biomimicry and biomaterials: Exploring interactivity, connectivity and smartness in nature. In V. Rognoli & V. Ferraro (Eds.), *ICS materials: Interactive, connected, and smart materials*. Milan: Franco Angeli

Pollini, B., Pietroni, L., Mascitti, J., & Paciotti, D. (2021). Towards a new material culture: bio-inspired design, parametric modeling, material design, digital manufacture. In *XII National Assembly SITdA Design in the digital age: Technology, nature, culture*. Naples: University of Naples Federico II.

Why is the research needed?

Given the young age of biodesign, basic research is needed. Bio-based materials' sustainable features are the main trigger for designers' experimentation—even though biofabricated ones have yet to be analyzed through sustainability metrics. Filling this gap would encourage innovative future productions, fitting the requirements of the emerging circular economy.

Materials Experience 2. DOI: https://doi.org/10.1016/B978-0-12-819244-3.00008-9

Which aspects of "materials experience" does the research valorize?

The relationship with living materials enables unique experiences; a fundamental aspect to investigate is how the sense of care for living materials could bring sustainable behavior changes.

How will the research impact on designers?

Biofabricated materials, with their potential and limitations, require new approaches and methodologies for designers. Supporting experimental exploration and a strong multidisciplinarity will be the basis of these new design tools.

What outcomes have been achieved or are foreseen?

Biodesign is still missing an established vocabulary. A glossary is under definition, based on a taxonomy derived from the fundamental features of biofabricated materials emerging through the research. Guidelines to define and design life-enabling materials are under definition, while more research will be needed to frame biofabricated ones within sustainability metrics.

What is the next big challenge for the research area?

The change toward more sustainable productions must be supported by a radical vision and management of material within a design project. A transdisciplinary approach oriented to a more-than-human design practice needs further development.

Which publications most inspired or informed the research?

Myers, W. (2012), *Bio design: Nature science creativity*. London: Thames & Hudson.

Camere, S., & Karana, E. (2018). Fabricating materials from living organisms: An emerging design practice. *Journal of Cleaner Production*, 186, 570–584.

Gatto, G., & McCardle, J. (2019). Multispecies design and ethnographic practice: Following other-than-humans as a mode of exploring environmental issues. *Sustainability*, 2019, 5032.

Guillitte, O. (1995). Bioreceptivity: A new concept for building ecology studies. *Science of the Total Environment, 167*(1–3), 215–220. https://doi.org/10.1016/0048–9697(95)04582-L.

Acknowledgments

The research is funded by Politecnico's PhD scholarship. As of now part of the study has been nurtured by the MADE project (materialdesigners.org) and through participation at "Circular sPRINT: Additive Manufacturing to foster Circular Design" organized by Ghent University and the Doctoral School of BSE (Bioscience) Engineering (http://www.crafth.eu).

The ongoing research is partly shared on the observatory healing-materialities design https://healing-materialities.design/home/.

The title of the research is the same as a lecture held within the conference "Caring Matters," organized within the project TAKING CARE by the Research Center For Material Culture, where I presented the ongoing research in a workshop session named "Healing Materialities," defining the potentialities of biofabricated materials that have emerged so far under this perspective.

FIGURE 1
Sample of bacterial cellulose grown by the author. © *Author.*

FIGURE 2
Demonstration of the recipe of a bioplastic during the MADE workshop given by the author. © *Author.*

FIGURE 3
Prototyping of ceramic 3D-printed tiles for the Circular sPRINT, designed by the author for the development of bioreceptive circular materials and objects, printed in the Biofablab of Ghent University. © *Author.*

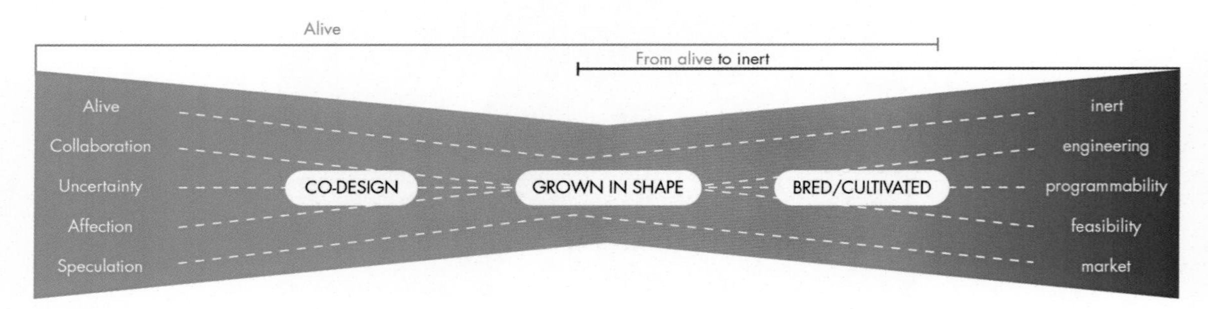

FIGURE 4
Taxonomic scale developed to analyze the case studies under three main aspects of the project: the speculative or feasible nature, the live/inert state of the organism, and its degree of involvement in the design phase.

PROJECT 14

Demonstrating a material making process through the cultivation of fungal growth

Dilan Ozkan
School of Architecture, Planning and Landscape, Newcastle University, Newcastle upon Tyne, United Kingdom

Supervisors: Prof. Dr. Martyn Dade-Robertson and Dr. Beate Christgen

Today, mycelium is used as a biomaterial in many different ways such as packaging, acoustic panels, wall insulation, bricks in buildings, textiles, and as a raw material in designed products including furniture. All these applications are realized through molding, which is an example of a constraining intervention, currently used in biodesign. In other words, cultivation is based on forming a predefined structure such as with a mold, where the organism's cells fill the negative space in a confined setting. For fungal material production, this process is very restrictive, reduces the degrees of freedom, and there is no room for emergence. Biomaterials lose their inherent potentials that come from owning a metabolism capable of material growth in an almost limitless way. Within my research, I am investigating fungi by posing the question: "how we can guide a living material [fungus] that has its own tendencies and cultivate it with minimum intervention?"

The first phase of my study involves experimentation by paying close attention to factors that might cause a difference in the behavior of mycelium, to understand its properties and nature. After having understood its behavior in response to different factors such as oxygen concentration, humidity, light intensity, temperature, and nutrient sources, the research will continue by guiding the growth of fruiting bodies. Integrating computational tools into the biofabrication process will aid designers to manipulate their living material robotically. Building a growth chamber regulated by sensors, algorithms, and finite state machines such as a fan, humidifier, light, and heater will help to parametrize and systematize the growth. In conclusion, designing my tools (robotic mold) in order to interact with the fungus and changing its morphology will help further manipulations of fungi for different intended purposes.

Stefanova, A., Arnardottir, T. H., Ozkan, D., & Lee, S. (2019). Approach to biologically made materials and advanced fabrication practices. In *Proceedings of the international conference on emerging technologies in architectural design* (pp. 193–200). Toronto: Ryerson University.

Why is the research needed?

This research demonstrates a new biological fabrication method and moves beyond the traditional ways of producing biomaterials. Robotic molding helps standardized production with its interactive and adaptive features.

Which aspects of "materials experience" does the research valorize?

The research furthers the implementation of morphology—the communication and dynamic interaction between a crafter and a living material—by which an

Materials Experience 2. DOI: https://doi.org/10.1016/B978-0-12-819244-3.00011-9

organism changes its form in response to modifications made by the designer.

How will the research impact on designers?

Building a dialog with an organism demonstrates a new kind of material making process, contributing to the architectural legacy of biomaterial tectonics and enhancing existing practices through opportunities for innovation.

What outcomes have been achieved or are foreseen?

In the initial phases, the parameters that manipulate the morphogenesis of fungus have been illustrated. According to these parameters, an automated device has been proposed and developed capable of influencing the growth of fungi remotely. This device facilitates growth by sensing, thinking, and acting, therefore it can be called a robot.

What is the next big challenge for the research area?

Fungi offer a vast array of functions—using it as a host material to form a symbiosis with other organisms or materials can be seen as the next challenge for the research.

Which publications most inspired or informed the research?

Adamatzky, A., Ayres, P., Belotti, G., & Wosten, H. (2019). Fungal architecture (pp. 1–19). Unpublished article submitted to *International Journal of Unconventional Computing*. <https://arxiv.org/abs/1912.13262> Accessed 05.06.20.

Hamann, H., Soorati, M. D., Heinrich, M. K., Hofstadler, D.N., Kuksin, I., Veenstra, F., Wahby, M., Nielsen, S.A., Risi, S., Skrzypczak, T., Zahadat, P., Wojtaszek, P., Støy, K., Schmickl, T., Kernbach, S., Ayres, P. (2017). Flora robotica: An architectural system combining living natural plants and distributed robots (pp. 1–16). Unpublished article. <http://arxiv.org/abs/1709.04291> Accessed 05.06.20.

Zolotovsky, K. (2017). *Guided growth: Design and computation of biologically active materials* [PhD thesis, Massachusetts Institute of Technology, Cambridge]. <https://dspace.mit.edu/handle/1721.1/113925> Accessed 05.06.20.

Ingold, T. (2000). Making things, growing plants, raising animals and bringing up children. In *The perception of the environment: Essays on livelihood, dwelling and skill* (pp. 77–88). London: Routledge.

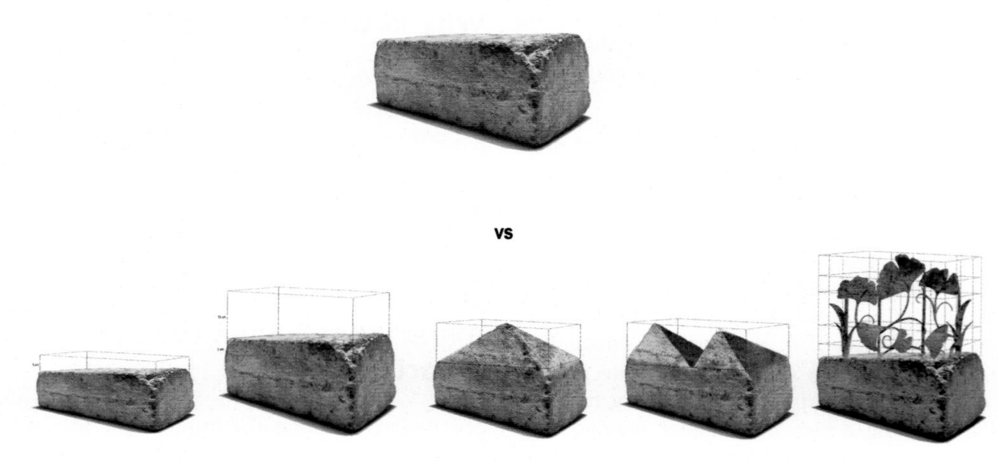

FIGURE 1
Possibilities of mycelium objects. © *Author.*

FIGURE 2
Going beyond the brick. © *Author.*

FIGURE 3
Mycelium growth in nutritionally heterogeneous environments. © *Author.*

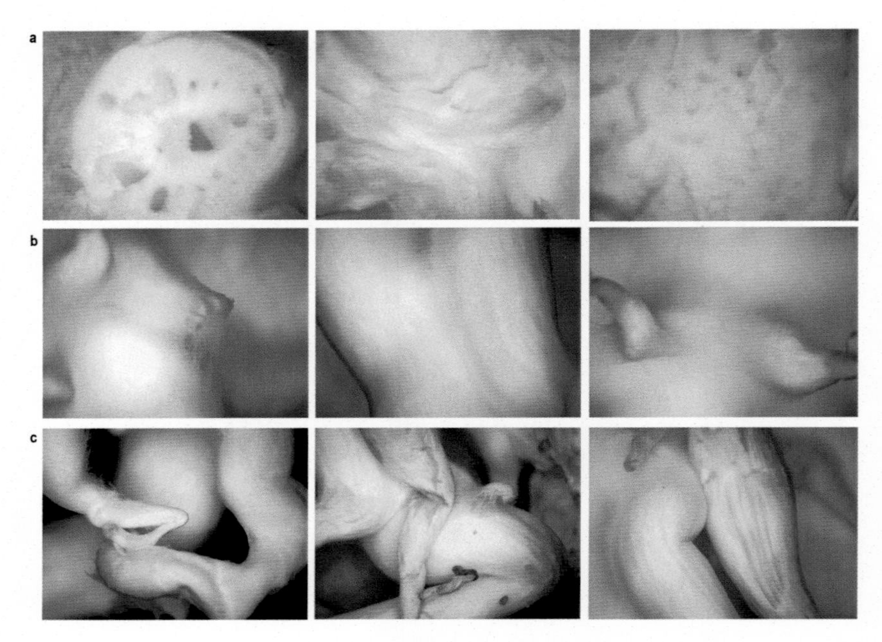

FIGURE 4

Fruiting bodies of *P. Ostreatus* (oyster mushroom): (A) Horizontal cut of fruiting bodies, (B) hyphae aggregate, and (C) vertical cut of fruiting bodies. © *Author.*

Malfunction, maintenance, and materials

Alexandra Karakas
Doctoral School of Philosophy, Eotvos Lorand University, Budapest, Hungary

Supervisor: Assoc. Prof. Mihály Héder

Most studies in the field of both philosophy of technology and design theory tend to focus on the problem of function and its varied nature, but there has been little discussion on malfunction. I argue that not only function but also malfunction is the fundamental characteristic of technical artifacts, since this concept only applies to human-made objects and it establishes the judgments attached to particular artifacts as well. Malfunction as such can refer to several different anomalous scenarios, including poor design and manufacturing mistakes. In all cases, malfunction indicates an artifact's inability to perform its intended function. In my research, I argue that malfunction is embedded in the material of an artifact, and that malfunction, as a spatiotemporal quality, defines even more the meaning and lifespan of objects than just functional features. Achieving a better understanding of artifacts, and specifically the relation between malfunction and materials, is not only a philosophical question, but a path to practical consequences. In my research the emphasis is on how malfunction is embedded in the material of a particular object—with a focus on aluminum—and how the combination of material and malfunction relates to social and historical circumstances.

The aim of the research is to construct a taxonomy of malfunction, focusing on the role of materiality. What happens when a material is not suitable for a particular design, it wears out too fast, or the material cannot be maintained? How can designers manipulate consumption through the choice of certain materials? How can malfunction be embedded in the material of artifacts, and how do certain materials determine the reception of designed objects? The scope of the research covers malfunction both from a design research perspective and from an ethical perspective.

Karakas, A. (2019). Maintenance, function, and malfunction in technology. *Információs Társadalom*, 2019/4, 1–12.

Why is the research needed?

Even though malfunction and failure are crucial features of technology and design, their role and status in particular devices and greater systems are still unclear. The purpose of the research is to give a better understanding of malfunction in design through the lens of materiality and to construct a taxonomy of different notions of malfunction.

Which aspects of "materials experience" does the research valorize?

The research emphasizes the connection between materiality and malfunction through the material experiences of aluminum.

How will the research impact on designers?

As well as philosophical arguments, the research seeks to outline the practical consequences of malfunction in design and technology.

Materials Experience 2. DOI: https://doi.org/10.1016/B978-0-12-819244-3.00005-3

What outcomes have been achieved or are foreseen?

Malfunction is increasingly a major problem in technology and design. Through the research, a theoretical framework for malfunction is proposed that analyses the notion of failure in order to offer guidelines to designers and to initiate a philosophical investigation about malfunction.

What is the next big challenge for the research area?

Currently the research focuses on physical objects, but software errors are becoming a major issue—the next step can be to analyze both the materiality of software and to develop a taxonomy for software malfunction.

Which publications most inspired or informed the research?

Kroes, P. (2010). Engineering and the dual nature of technical artefacts. *Cambridge Journal of Economics, 34* (1), 51–62.

Karana, E. (2009). *Meanings of materials* [PhD thesis, Delft University of Technology, Delft].

Preston, B. (2009). Philosophical theories of artifact function. In A. Meijers (Ed.), *Handbook of philosophy of technology and engineering sciences* (pp. 213–233). Burlington: North Holland.

Franssen, M. (2006). The normativity of artefacts. *Studies in History and Philosophy of Science Part A, 37* (1), 42–57.

FIGURE 1
Women using aluminum cookware (Hungary, 1970).
Creative Commons 3.0, CC BY-SA 3.0.

FIGURE 2
Flea market with aluminum artifacts (Hungary, 1916).
Creative Commons 3.0, CC BY-SA 3.0.

FIGURE 3
Cookware shop (Hungary, 1959). *Creative Commons 3.0, CC BY-SA 3.0.*

FIGURE 4
Open fire cooking (Hungary, 1942). *Creative Commons 3.0, CC BY-SA 3.0.*

Open-Ended Design: how to intentionally support change by designing with imperfection

Francesca Ostuzzi
Department of Industrial Systems Engineering and Product Design, Ghent University (Campus Kortrijk), Ghent, Belgium

Supervisors: Prof. Dr. Ing. Jelle Saldien and Prof. Dr. Ing. Jan Detand

My work is explorative. It adopts qualitative research methods focusing on the design practice and creation of knowledge "through design." My thesis articulated six studies, as follows.

Study 0 (observation) analyzed more than 100 industrial products that embed ingenious and rather intuitive ways to meaningfully embrace change in design. Examples included products that are designed with imperfections that make them more resilient when facing out-of-control possibilities of breakage, aging, etc. Specifically, in the arena of product–environment interaction (inhabited by human and nonhuman actors), materials have a dynamic nature that conveys information and elicits new meanings. Materials' imperfections allow emergence, crucial for the open-endedness of products.

Studies 1–4 (anticipation) observed the creation of more than 70 original and intentional Open-Ended Design outcomes. The aim was to understand how we can create Open-Ended designs (e.g., what should be left open and what strictly defined?) and how we can effectively distribute them.

Study 5 reached a closure, unifying all my understanding into a proposal of an *Open-Ended Design approach* intended to support designers. The approach is not rigid in its structure and does not aim to be prescriptive. It is structured as an iterative process where the designer is asked to constantly switch from observing the dynamic nature of reality, to anticipating it, through a design act. To this end, I formalized a set of "lenses" that prompt different perspectives for the designer during both observation and anticipation phases. What changes? Why? When? Where? How much? How fast? With which goal? Is the change reversible?

Designers are no longer asked to think only with reference to standardization where design solutions are given and remain "perfect" and stable in time. I challenge designers to acknowledge the dynamicity of design outcomes, which are time- and context-related, out-of-control and undisciplined, but nevertheless capable of anticipation (although not "prediction") by an intentional design act.

Ostuzzi, F., De Couvreur, L., Detand, J., & Saldien, J. (2017). From design for one to open-ended design: Experiments on understanding how to open-up contextual design solutions. *The Design Journal*, 20 (Suppl. 1: Proceedings of the 12th European Academy of Design conference—Design For Next), 3873–3883.

Ostuzzi, F., Rognoli, V., Saldien, J., & Levi, M. (2015). +TUO project: Low cost 3D printers as helpful tool for small communities with rheumatic diseases. *Rapid Prototyping Journal, 21*(5), 491–505.

Why is the research needed?

Products are anything but static, predictable, or under control. In fact, they are dynamically subject to

Materials Experience 2. DOI: https://doi.org/10.1016/B978-0-12-819244-3.00025-9

constant transformations. The motivation for this research was a need to properly describe this "phenomenology of change" in relation to designed products. What is its role and potential value? How can designers intentionally embrace the (often unavoidable) process of change?

Which aspects of "materials experience" does the research valorize?

Moving on from values attributed to imperfect material esthetics (e.g., accumulation of usage traces or aging processes), the research provides an approach through which dynamic materials' experiences become outcomes of intentional design acts.

How will the research impact on designers?

Open-Ended Design aims at triggering reflections on the way we design and interact with products that is, ultimately, the way we participate in creating more or less sustainable futures.

What outcomes have been achieved or are foreseen?

Open-Ended Design is proposed as an approach to observe (post-factum) and generate (ante-factum) the conditions under which changes in products become less disruptive or sometimes even desirable. Open-Ended Design outcomes are suboptimal, error-friendly, "wabi sabi," context-dependent, and characterized by their inner flexibility due to the voluntary incomplete definition of their features.

What is the next big challenge for the research area?

The emergent topics of biological materials in design, growing materials, and, more generally, the Circular Economy strongly relate with Open-Ended Design—how can we support these paradigm shifts using Open-Ended Design?

Which publications most inspired or informed the research?

Dubberly, H., & Pangaro, P. (2015). Cybernetics and design: Conversations for action. *Cybernetics and Human Knowing, 22*(3), 73–82.

Nelson, H., & Stolterman, E. (2012). *The design way: Intentional change in an unpredictable world.* London: MIT Press.

Redström, J. (2008). RE: Definitions of use. *Design Studies, 29*(4), 410–423. 10.1016/j.destud.2008.05.001.

Wakkary, R., & Maestri, L. (2008). Aspects of everyday design: Resourcefulness, adaptation, and emergence. *International Journal of Human-Computer Interaction, 24*(5), 478–491. 10.1080/10447310802142276.

FIGURE 1

Examples of dynamic product uses, due to changes at the users' interpretation level. © *Author.*

FIGURE 2
Examples of dynamic product appearances and uses, due to changes at the materials level. © *Author.*

FIGURE 3
Reading feedback from the traces in a product: an Open-Ended house that triggers reappropriation. © *Elemental (elementalchile.cl)*

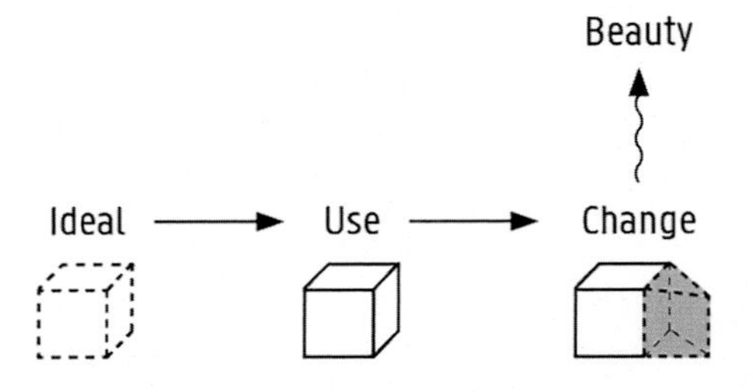

FIGURE 4
Beauty is not created with the product but emerges in time and through use. © *Author.*

Material information platform for designing environmentally friendly products

Indji Selim
Department of Industrial Design, Ss. Cyril and Methodious University Skopje, Skopje, Republic of North Macedonia

Supervisor: Prof. Dr. Tatjana Kandikjan

The industrial designer creates a product by balancing between consumers' needs and manufacturers' requirements, in which selecting the most suitable material for that particular product is an inherent task of the designer's process. However, the unfamiliarity of attributes of recently discovered environmentally friendly materials (EFM) presents a challenge to designers in the material selection process. My research focuses on the role of the industrial designer within the scope of EFM selection.

The material information platform (MIP) that I propose aims to emulate the human decision-making process, in this case, the decision-making process for product-driven design (PDD) and its evolution throughout the material selection process, focused on EFM. For this purpose, I conducted research to support a five-phase PDD materials selection method, visualizing the process and the material decisions through the example of a single-use water bottle.

In the first phase, I used a multiattribute decision-making (MADM) method to define fixed and alternate product criteria. Product attributes mutually required by the designer, consumer, and manufacturer were denoted as fixed product criteria, while the remaining attributes that were outside the mutual cross-section were denoted as alternate product criteria. Based on the associative description concept, I mapped the product attributes onto the attributes of a generic material. By considering whether the mappings related to fixed or to alternate material criteria, I developed a material criteria chart to provide the decision-maker (designer) with an evaluative approach toward materials selection, in favor of browsing the most suitable materials. Based on the valued alternate material criteria, I assembled four different EFM alternatives for the single-use water bottle using different online sources. In the final phase, I tested the suitability of each material by presenting a table of five inherent attributes. Having the PDD method in perspective, the outcome of the research is to develop an assistive tool referred to as MIP by accommodating further material groups and product-related attributes.

Selim, I., Lazarevska, A. M., Mladenovska, D., Kandikjan, T., & Sidorenko, S. (2020). Identifying material attributes for designing biodegradable products. In I. Karabegovic (Ed.), *New technologies, development and application* (pp. 633–639). Cham: Springer.

Selim, I., Lazarevska, A. M., Kandikjan, T., & Sidorenko, S. (2019). Multi-attribute material information platform. In N. Börekçi, D. Jones, F. Korkut, & D. Özgen Koçyıldırım (Eds.), *Proceedings of DRS Learn X Design 2019—fifth international conference for design education researchers* (pp. 531–541). Ankara: Middle East Technical University.

Why is the research needed?

The aim of this research is to develop a MIP as an assistive tool that collects and aggregates product requirements—based on both consumers' and manufacturers' preferences—and thus offers designers substantial information regarding EFM and their features. The assistive tool suggests an optimal set of materials that matches the predefined product requirements.

Materials Experience 2. DOI: https://doi.org/10.1016/B978-0-12-819244-3.00035-1

Which aspects of "materials experience" does the research valorize?

Based on the knowledge incorporated in previously produced products (i.e., gathering and categorizing significant materials experience), the MIP provides assistive guidance to the designer, throughout the EFM selection process.

How will the research impact on designers?

The MIP saves time, is able to keep updated with constantly changing information, and gives more options and opportunities to designers otherwise reliant on personal experience.

What outcomes have been achieved or are foreseen?

So far, various EFM are accumulated and aggregated according to the PDD method. The next step is to cross-reference the EFM-related data (attributes) and enlarge the MIP with additional product examples. Progressively, suitable MADM tools and machine learning algorithms will be proposed and possibly incorporated in line with the refined MIP architecture.

What is the next big challenge for the research area?

The method developed through this research refers to EFM alternatives—however, the creation of the refined MIP architecture and the incorporation of MADM tools and machine learning algorithms is the next big challenge.

Which publications most inspired or informed the research?

Karana, E., Barati, B., Rognoli, V. & Zeeuw van der Laan, A. (2015). Material driven design (MDD): A method to design for material experiences. *International Journal of Design, 9*(2), 35–54.

Karana, E. (2009). *Meanings of materials* [PhD thesis, Delft University of Technology, Delft].

Ashby, M., & Johnson, K. (2002). *Materials and design: The art and science of material selection in product design*. Oxford: Butterworth-Heinemann.

Keeney, R. (1982). Decision analysis: An overview, *Operations Research, 30*(5), 803–838.

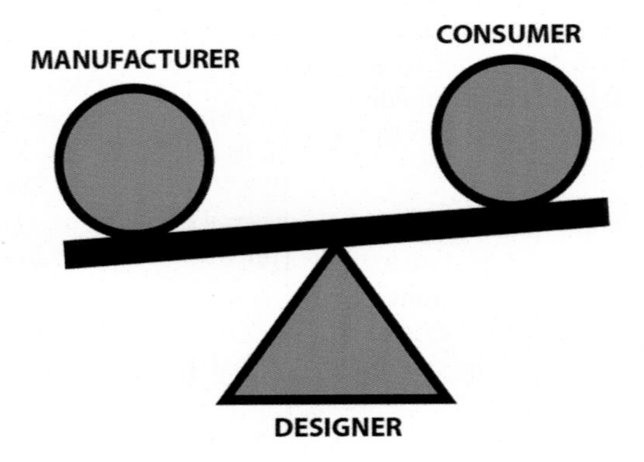

FIGURE 1
The industrial designer's pivot and balancing role. © *I. Selim, M. A. Lazarevska, D. Mladenovska, T. Kandikjan.*

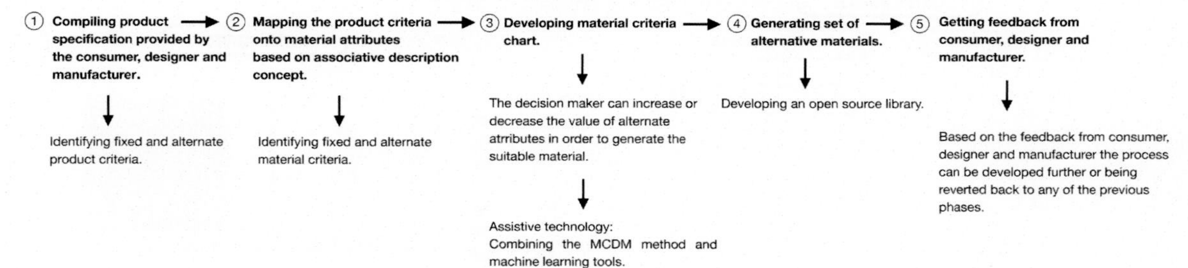

FIGURE 2

Material selection process based on product-driven design (PDD). © *I. Selim, M. A. Lazarevska, T. Kandikjan, S. Sidorenko.*

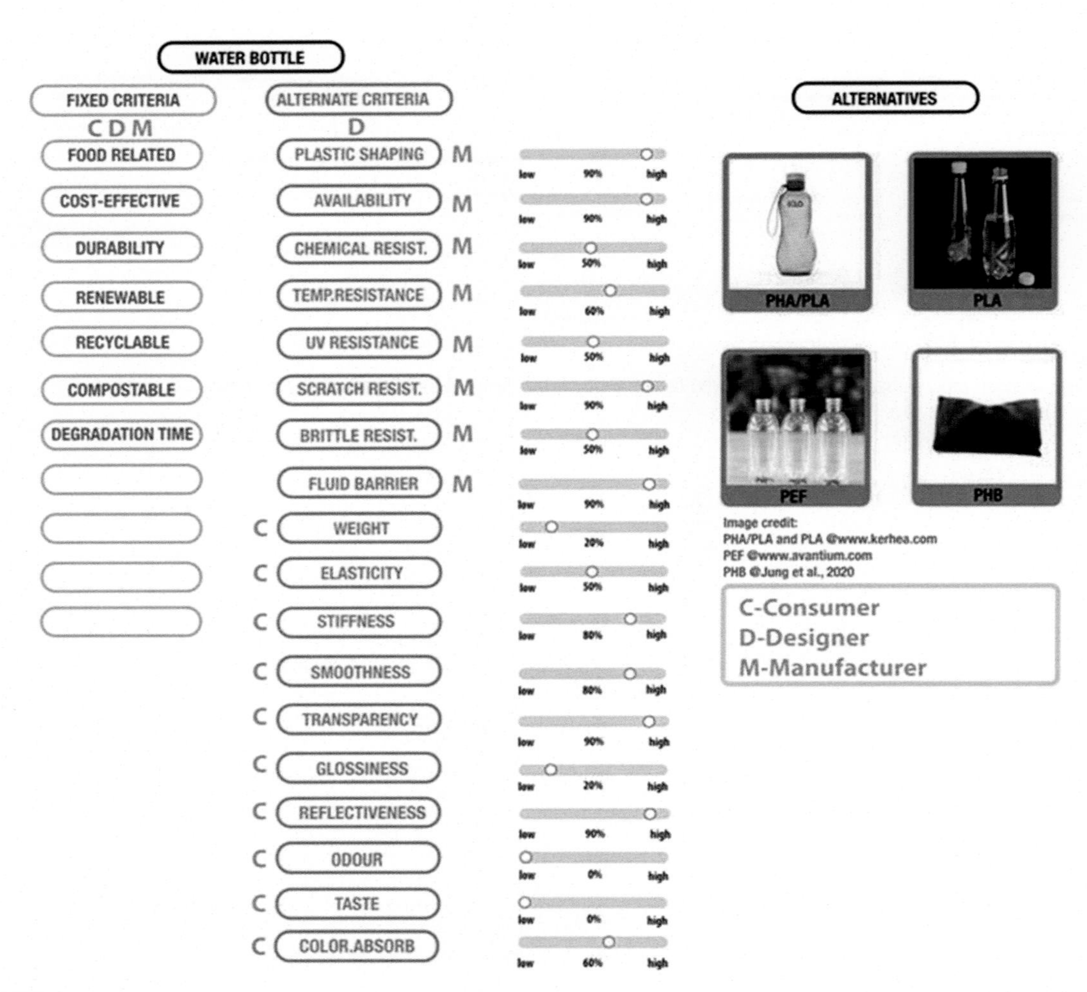

FIGURE 3

Categorization of material criteria (fixed and alternate) accompanied by suitably matched material set. © *I. Selim, M. A. Lazarevska, T. Kandikjan, S. Sidorenko.*

Image credit:
PLA @www.kerhea.com

GENERAL ATTRIBUTES

Price: 2 eur.per kg
Usage: Medical appliances, fibers and textiles, packaging and service ware, electronics and agricultural applications.
Processes:
Blow-molding, injection molding, extrusion, cast film extrusion, fiber spinning, thermoforming and 3D priniting.
Ashby and Johnson (2002)

TECHNICAL ATTRIBUTES

Melt temp.: 157-170 °C
Injection molding temperature: 178-240 °C
Fire resistance: unknown
UV resistance: moderate
Scratch resistance: good
Weight: medium
Chemical resistance: moderate
-Good barrier properties
-Thermally unstable
-Insoluble in water
Ashby and Johnson (2002)
materialdistrict (2018)

AESTHETICAL ATT.

Glossiness: glossy
Transparency: 50-100%
Structure: closed
Texture: smooth
Hardness: medium
Temperature: medium
Acoustics: poor
Odour: none
(materialdistrict, 2018)

ENVIRONMENTAL ATT.

Biodegradable
Compostable in industrial composters
Recyclable
Landfilling
Primary production:
Embodied energy: 49 -54 MJ/kg
CO_2 footprint: 3.4 -3.8 kg/kg
Water usage: 100-300 L/kg
Eco-indicator: 278 milli-points/kg
Processing
Polymer molding energy: 15.4-17 MJ/kg
Polymer molding CO_2: 1.15-1.27 kg/kg
Polymer extrusion energy 5.7-6.3 MJ/kg
Polymer extrusion CO_2: 0.43-0.47 kg/kg
End of life
Embodied energy, recycling: 33-40 MJ/kg
CO_2 footprint, recycling : 2.0 -2.4 kg/kg
Recycle fraction in current supply 0.5 -1 %
Heat of combustion 18.8 -20.1 MJ/kg
Combustion CO_2: 1.8 -20.1 MJ/kg
Recycle mark: 1.8-1.9 kg/kg
(Ashby, 2012)

USER EXPERIENCE

-Sun chips brand withdraw their PLA packaging because customers didn't like the noisy packaging.
-It is brittle
Kyle VanHemert (2010)

-It has shiny and smoothy appearance and it smells sweet when printed in 3d printer.
–Can deform because of heat (like a cassette in a car)
Ramon (2013)

FIGURE 4

Categorizing the material attributes for one of the matched materials. © *I. Selim, M. A. Lazarevska, T. Kandikjan, S. Sidorenko.*

Material education in design: engaging material experimentation and speculation

Ziyu Zhou
Design Department, Politecnico di Milano, Milan, Italy

Supervisor: Assoc. Prof. Dr. Valentina Rognoli
Co-supervisor: Assoc. Prof. Dr. Manuela Celi

In the last 15 years, we have witnessed the growing importance of the expressive sensorial dimension of materials with the birth of the materials experience concept, and this has influenced materials for design education. Today in some schools around the world, students are being encouraged to experiment with materials and to use, more or less consciously, practical approaches to manipulate those materials by designing their qualities and identities.

My research started with an investigation on the world of materials education, to understand its evolution and the fundamental aspects essential for the contemporary training of the designer. How can we implement and strengthen the designer's knowledge in the field of materials? What tools and methods could we give to students to help them explore materials and technologies and to facilitate their intervention or contribution to the evolution of the material world?

To compile an exhaustive study of state of the art (Phase 1), using the tools for desk research, I browsed many materials' courses from design schools all over the world and highlighted the essential aspects of the thinking and practice of materials and design, as well as the courses' relevant pedagogical approaches. Afterwards, I conducted ethnographic research (Phase 2), by including the construction of experience, the practical approach with materials, the blending of personal interests and cultural backgrounds, the search for alternative resources, and the critical and speculative attitudes toward materials.

At the moment, I have advanced the assumption (Phase 3) of possible educational models on the aspects of thought, practice, and communication. `I formulated an assumption around materials questioning, experimentation, and integration. Thus, working toward the implementation of a new materials educational model in seminars or design courses, I am looking forward to observing and verifying the assumption through the application of my ideas in a real class.

Zhou, Z., & Rognoli, V. (2019). Material education in design: from literature review to rethinking. In: *Proceedings of the 5th international conference for design education researchers (DRS Learn-X 2019)* (pp. 111–119). Ankara: Middle East Technical University, Department of Industrial Design.

Zhou, Z., Rognoli, V., & Ayala-Garcia, C. (2018). Educating designers through Materials Club. In: *Proceedings of the 4th international conference on higher education advances (HEAD'18)* (pp. 1367–1375). València: Editorial Universitat Politècnica de València.

Why is the research needed?

The main aim of the research is to improve materials education in design and make it coherent with the contemporary world of materials, which is continuously changing and evolving. The particular focus is on the possibility of including multidisciplinary experimentation in the designer's education alongside a critical and speculative approach.

Materials Experience 2. DOI: https://doi.org/10.1016/B978-0-12-819244-3.00024-7

Which aspects of "materials experience" does the research valorize?

The research enhances the materials experience concept in all its aspects, because future designers have to learn to manage all its different nuances and variations from an educational perspective.

How will the research impact on designers?

By elevating awareness of the multiple roles of materials, designers will improve their skills and competencies to choose and design the right materials for their projects.

What outcomes have been achieved or are foreseen?

The aimed outcomes are primarily pedagogical guidelines to outline a customizable course or educational program that can enhance experimentation and speculation with materials. The materials, and practices around the materials, can be promoters of new visions and give rise to reasoning on the material qualities and possible and preferable applications.

What is the next big challenge for the research area?

The next big challenge is for materials-centered education to become an internationally recognized research area, in which design schools can share similar and complementary educational paths.

Which publications most inspired or informed the research?

Dunne, A., & Raby, F. (2016). *Speculative everything: Design, fiction, and social dreaming*. Cambridge: MIT Press Ltd.

Drazin, A., & Küchler, S. (Eds.) (2015). *The social life of materials: Studies in materials and society*. London: Bloomsbury.

Karana, E., Pedgley, O., & Rognoli, V. (Eds.) (2014). *Materials experience: Fundamentals of materials and design*. Oxford: Butterworth-Heinemann.

Martinez, S. L., & Stager, G. (2013). *Invent to learn: Making, tinkering, and engineering in the classroom*. Torrance: Constructing Modern Knowledge.

Acknowledgment

This research is funded by the China Scholarship Council.

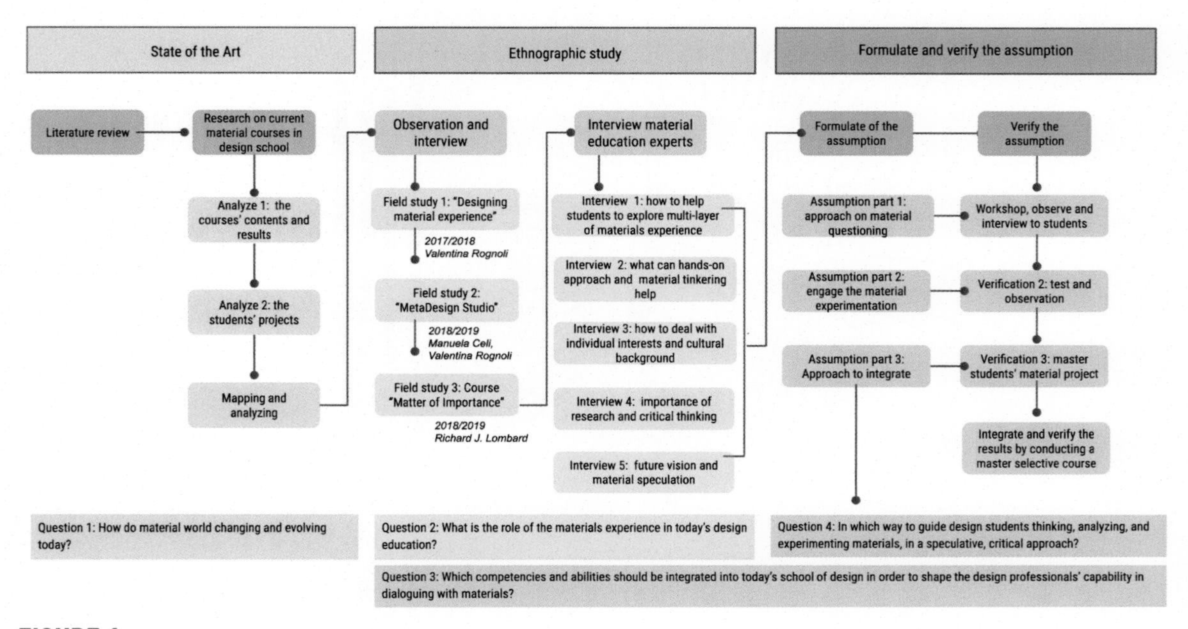

FIGURE 1
Research methodology framework. © *Author.*

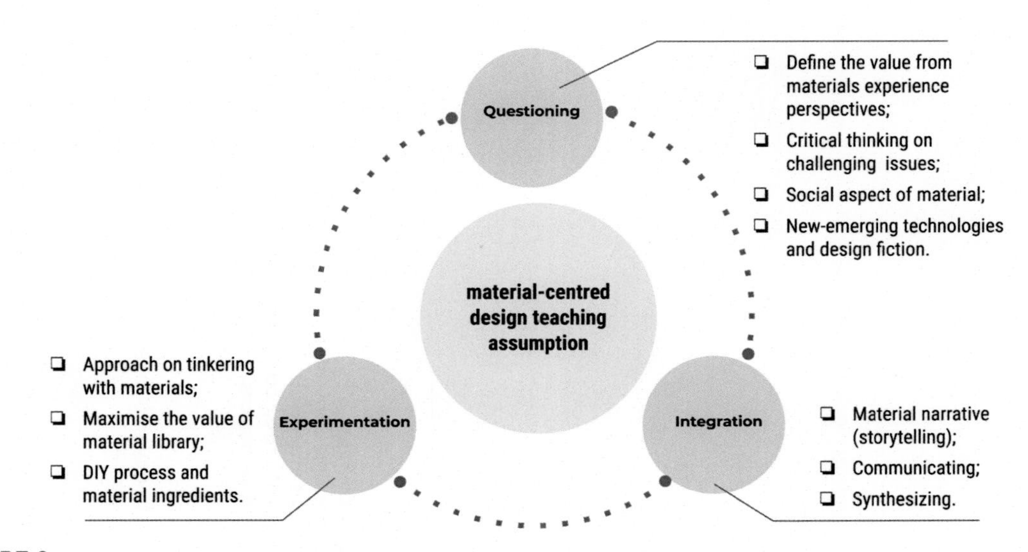

FIGURE 2
Framework: material-centered design teaching assumption. © *Author.*

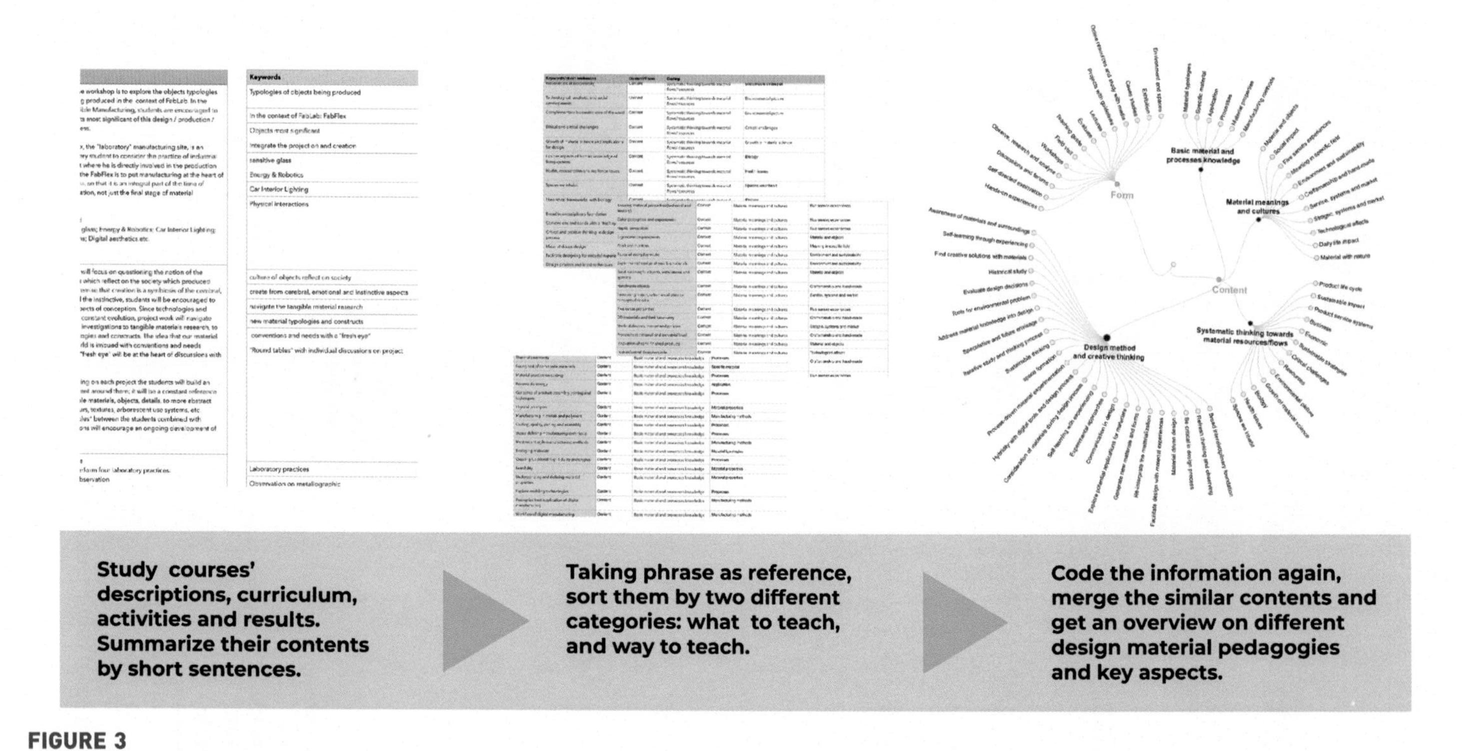

FIGURE 3

Research progress: analysis of material courses, screenshot in 2019. © *Author.*

CHAPTER 7

A renewed recognition of the materiality of design in a circular economy: the case of bio-based plastics

Conny Bakker, Ruud Balkenende
Faculty of Industrial Design Engineering, Delft University of Technology, Delft, Netherlands

7.1 Introduction

In 1989 Ezio Manzini wrote *The material of invention*. In the book, he argues that designers have been freed from the constraints that materials pose. In Manzini's vision, materials can now be engineered to fit the needs of the designer, instead of the other way around. These "materials made to order" (p. 37) herald a new age where material selection is no longer needed and where the choice of materials becomes, in fact, design. More than two decades later, Graedel et al. (2015) tentatively raise the alarm, as they see an escalating trend in material consumption, with today's technologies employing nearly every material in the periodic table of elements. They note that this had led to increasing materials complexity, while recycling rates remain low, the associated environmental impacts of virgin materials production high, and options for material substitution limited. The use of so many materials, they state, "is a carefully calculated effort to achieve increasingly high performance in products simple to complex" (p. 6295). In other words, it is an act of design.

We have thus become highly dependent upon a wide range of materials for our modern technologies, and this has come at a cost. According to the United Nations, "the extraction and processing of materials, fuels and food make up about half of total global greenhouse gas emissions and more than 90% of biodiversity loss and water stress" (IRP, 2019). In addition, with 55% of materials currently being recycled in the European Union (Eurostat, 2019), but with recycling rates of 30 out of the 60 different metals in use at less than 1% (United Nations, 2011), we need to have a long hard look at the way we design products and, therefore, select and use materials.

In particular, designers need to better understand the (impacts of the) origin and composition of the materials they design with, and, perhaps even more importantly, their recovery and end-of-life options. This makes the circular

Materials Experience 2. DOI: https://doi.org/10.1016/B978-0-12-819244-3.00020-X

economy such an interesting concept for designers. The circular economy is defined as an industrial system that is regenerative and restorative by design (EMF, 2013). One of the aims of a circular economy is to attain high-value and high-quality material cycles (Korhonen et al., 2018) where products and materials can loop back into the economy without much (or any) loss of quality and value, or where products and materials can be safely "fed back" into the biosphere to be regenerated. If we would have a well-functioning circular economy, we would be able to recover and reuse products and materials many times over, which would vastly reduce greenhouse gas emissions for the production of virgin materials, as well as a lot of unnecessary waste. In other words, a circular economy would transform the way we make, use, and recover products and materials.

In this chapter we intend to sketch the opportunities and challenges of designing products and materials that fit a circular economy. We also want to take a longitudinal perspective with a focus on managing vital resources for the long term. We thus take as our starting point the phase-out of fossil fuels, which is inevitable given the escalating climate crisis. This chapter will therefore focus on bio-based materials in a circular economy. Bio-based materials are made from renewable resources. While wood, leather, etc., are well-known bio-based materials, this chapter will look at bio-based plastics in particular, because these can substitute for fossil-based plastics and contribute to the phase-out of fossil fuels and because plastics are used at a large scale in industrial product design.

The main objective of this chapter is to explore the opportunities and challenges of designing with bio-based plastics in a circular economy. We will show that the often-used "butterfly" model of the circular economy, which distinguishes between a biocycle (biological cycle) and a technocycle (technical cycle), is not very useful when we consider bio-based plastics from a design perspective. This leads us to propose an alternative framing of the circular economy with a limited number of recovery pathways. Based on this framing, we will show that bio-based plastics can provide new combinations of properties for new applications, as well as the straightforward replacement of existing fossil-based polymers. It follows that designers will be given new opportunities and will need to deal with new challenges as these complex and exciting bio-based materials enter the market in increasing numbers.

7.2 Bio-based and biodegradable plastics: a short review

While most designers have a working knowledge of natural and renewable materials such as wood, cork, cotton, and wool, they remain relatively ignorant of the material opportunities offered by bio-based plastics. Following

Lambert & Wagner (2017), we define bio-based plastics as plastics derived from renewable feedstock. It is important to realize that a plastic produced from renewable feedstock is not necessarily biodegradable (biodegradable plastics can decompose through the action of microorganisms) and that plastics produced from fossil resources *can* be biodegradable. In this chapter we will talk about bio-based plastics and not bio-plastics, because the term bio-plastics are confusingly used for both bio-based plastics and biodegradable plastics. What follows is a short explainer and state-of-the-art review of the field of bio-based and biodegradable plastics.

7.2.1 Bio-based plastics

The bio-based plastics we discuss in this chapter are made from well-defined monomers obtained after processing of (parts of) plants or microorganisms such as algae. Often-used examples of bio-based plastics that can be used for engineering applications are PLA (polylactic acid) and PHA (polyhydroxyalkanoate). Common polymers like PE (polyethylene) and PP (polypropylene), that are currently mainly associated with fossil fuels, can also be formed upon further chemical processing of bio-based molecules (often via a bioethanol route). This results in bio-PE and bio-PP, also referred to as "drop-in" polymers, because they have the same properties as their fossil-based equivalents. Even partially bio-based polymers exist. Bio-polyethylene terephthalate (e.g., used in the Coca Cola "Plant bottle") is an example, in which one of the constituting monomers is bio-ethylene-glycol, while the other monomer is terephthalic acid made from fossil sources, implying that bio-PET consists of 30% plant-based materials (Sheldon, 2014). Polymers that directly use biomass such as chitin, cellulose, and starch are all biodegradable, but due to their properties they are less suited for engineering applications. See Table 7.1 for an overview.

The production of bio-based plastics has been criticized for competing with resources for food production. The "first-generation" feedstocks that used, for instance, edible seeds or corn (Sheldon, 2014) have sparked the development of

Table 7.1 Examples of bio-based and petro-based polymers. The table shows both biodegradable and nonbiodegradable examples.

	Petrochemical	Partially bio-based	Bio-based
Not biodegradable	PE, PP, PET, PS, PVC	Bio-PET ("drop-in")	Bio-PE ("drop-in") Vulcanized polyisoprene
Biodegradable	PBAT, PBS(A), PCL	Blends with starch	PLA, PHA, cellophane, cellulose, starch

PBAT, *Polybutylene adipate terephthalate;* PBS(A), *polybutylene succinate/adipate;* PCL, *polycaprolactone;* PHA, *polyhydroxyalkanoate;* PS, *polystyrene;* PVC, *polyvinyl chloride.*
Source: Adapted from Molenveld, K., & Van den Oever, M. (2014). Catalogus biobased verpakkingen. Wageningen UR Food & Biobased Research. ISBN 978-94-6173-704-5.

second-, third-, and even fourth-generation biostrategies. These use nonedible, fast-growing crops or agricultural by-products (second generation), algae biomass (third generation), or, as a new development on the horizon, "fourth generation" genetically modified algae and other microbes as feedstock (Azmah et al., 2016). In deconstructing and synthesizing raw biomaterials (feedstock) into chemicals for polymer production, there is a wide range of industrial processing techniques involved, which will not be discussed here. Just like fossil-based polymers, bio-based polymers will need additives such as antioxidants, plasticizers, colorants, and fillers to adapt their physical properties and improve their suitability for a particular use. According to Lambert & Wagner (2017), "because different manufacturers will add different additives at various concentrations, materials made from the same polymer should not be considered chemically identical." This may hamper their biodegradability and/or recyclability.

7.2.2 Biodegradable plastics

Biodegradable plastic is often confused with bio-based plastic. The term "bio-based" provides information regarding the renewable origin of the carbon source used for manufacturing the plastic. In contrast, "biodegradable" provides information on a potential degradation mechanism, which is completely unrelated to the carbon source. Biodegradable plastics can be made from fossil fuels as well as bio-based sources (see Table 7.1 for examples). Biodegradation thus does not affect the production phase of a material, but its disposal phase, its "degradation pathway."

For designers it is important to realize that when talking about biodegradability, there are a number of degradation pathways that all tend to be captured under the term biodegradability, but which are however very different and not all of them environmentally benign. Lambert & Wagner (2017) distinguish between degradation, biodegradation, and composting.

- Degradation is a process by which a material disintegrates. It can, for instance, take place as a result of natural daylight (photodegradation of plastics) or because special additives were added that help a plastic break down into small fragments over time (oxodegradable plastics). Oxodegradation of plastics can lead to the accelerated formation of microplastics.
- Biodegradation results from the action of naturally occurring microorganisms such as bacteria, fungi, and algae. Biodegradation results in the lowering of the molecular masses of the macromolecules that form the plastic; and in the case of "ultimate" biodegradation, in the complete reduction to simple molecules like carbon dioxide and water.
- Composting is a form of biodegradation that results in carbon dioxide and compost (water and biomass), which contains valuable nutrients

and is used as a soil improver. Composting is often misunderstood as a process that happens "naturally" when a biodegradable polymer ends up somewhere in the open environment. However, very few biodegradable polymers break down successfully at ambient temperatures in a reasonable timeframe. Most can only be composted under carefully controlled conditions, in industrial composting plants. The EU has a standard for determining compostability (EN 13432), which requires at least 90% disintegration after 12 weeks and 90% biodegradation in 6 months. A recent French standard has specified the requirements for home composting: at least 90% decomposition within 12 months at ambient temperature (NF T51-800).

The (bio)degradation of a polymer is affected by its structure, the additives used in the production of the material, and the exposure conditions such as humidity, temperature, pH, salinity, and the presence of sufficient oxygen (Lambert & Wagner, 2017). These need all to be taken into account when designing with such polymers. In the presence of oxygen, biodegradable plastic is converted into water and CO_2 by microorganisms. However, under anaerobic conditions, methane can be produced, which is a potent greenhouse gas.

7.2.3 Sustainability issues of bio-based and biodegradable polymers

From a sustainability perspective, the substitution of fossil-based plastics by bio-based plastics generally leads to lower nonrenewable energy use and greenhouse gas emissions (Mülhaupt, 2013). The greenhouse gas emission reduction, however, may be counteracted by direct and/or indirect land-use change (Oever et al., 2017). There are, however, some possibilities for a sustainable coproduction of biofuels, bio-based plastics, and food; this may serve as a stabilizer for food prices, providing farmers with more secure markets and contributing to social sustainability (ibid). Regarding the genetically modified organisms (GMOs) in fourth-generation biofeedstock, the potential release of GMO to the environment is considered a serious risk (Azmah et al., 2016).

In the case of oxodegradable plastics, the European Commission (2018) concluded that there is no evidence that the fragmentation of oxodegradable plastic in the open environment (land or sea) is sufficiently rapid to allow subsequent biodegradation to take place within a reasonable timeframe. Remaining plastic fragments and microplastics will lead to pollution of land and sea. Moreover, there are concerns within the recycling industry that the presence of oxodegradable plastic in a conventional plastic recycling system can lead to poor quality recyclate. It has led the European Commission to "start a process to restrict the use of oxo-plastics in the EU" (European Commission, 2018).

When it comes to biodegradation, a bio-based plastic will have to be pure, or safe, enough to prevent the quality of the final compost from declining, for example, because the plastic contains additives with heavy metals or fluorine. However, this is not sufficient. If we want to avoid plastic pollution, we need to assess the degradation process in a reasonable timeframe and not in terms of hundreds of years. This also implies that the degradation conditions should be taken into account. This is directly notable when looking into compostable polymers, which often can only be composted at elevated temperature under industrial conditions. PLA, for instance, will take hundreds of years to decompose under aquatic conditions (i.e., in the sea). It follows that designers need to be acutely aware of the kind of bio-based and biodegradable polymers they design with, in order to prevent unwanted negative sustainability effects.

7.3 Circular economy

The circular economy distinguishes between resource loops in the biocycle and the technocycle (EMF, 2013), as illustrated in Fig. 7.1. In the technocycle materials loop back into the economy without much (or any) loss of quality through reuse and industrial recycling. The ideal scenario in the technocycle

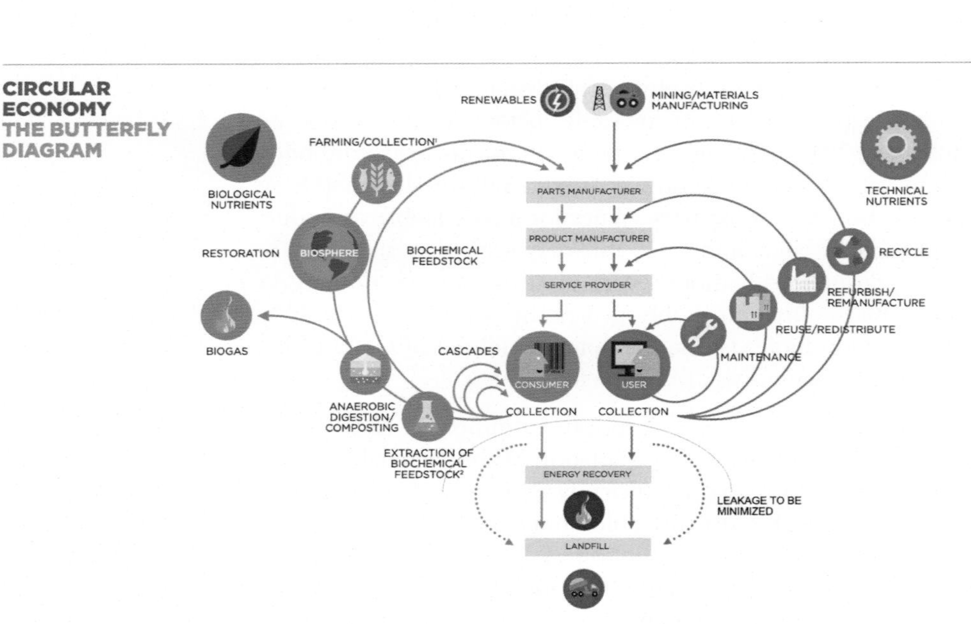

FIGURE 7.1
The butterfly diagram of the circular economy, showing the biocycle and technocycle. *Courtesy Ellen MacArthur Foundation <http://www.ellenmacarthurfoundation.org>*

is a slow but continuous cycling through "everlasting" loops. In the biocycle biodegradable materials are cycled back into the biosphere to be regenerated (either industrially or naturally). This should be done in consecutive cycles, named cascades, that see the materials lose quality until they are no longer fit for use. The ideal scenario is that materials in the biocycle cascade slowly and then enter the biosphere through (industrial) biodegradation.

7.3.1 Biocycle versus technocycle

We argue here that, from an industrial product design perspective, distinguishing between the biocycle and technocycle is not very helpful. Given the "ambiguous" nature of many materials (for instance, mixtures of biodegradable materials and nonbiodegradable additives) and the fact that bio-based materials are also used and recycled in industrial processes, the rigid distinction becomes confusing rather than helpful.

It is more useful to think in terms of options, rather than a dichotomy. Designers always need to take into account the recovery and end-of-life pathways of the specific products and materials they work with instead of worrying about whether, for instance, a bio-based polymer should go in the technocycle or the biocycle. Our new model gives an overview of recovery options and makes it easier to assess the risks of materials not ending up in the right recovery pathway (see Fig. 7.2).

Instead of the technocycle–biocycle division, we propose to visualize the circular economy as an economic system with a range of options, or recovery pathways, for products and materials.

- Maintenance and repair (product integrity focus)
- Direct reuse and redistribution (product integrity focus)
- Refurbishment and remanufacturing (product integrity focus)
- Recycling (material integrity focus)
- Molecular decomposition (carbon cycle focus)

The repair, redistribution, and refurbishment/remanufacturing loops of the "technocycle" remain similar to the technocycle in the butterfly diagram (see Fig. 7.1). The recycling and molecular decomposition loops change and are explained in more detail as follows.

7.3.2 Recycling of bio-based plastics (both nonbiodegradable and biodegradable)

Two types of recycling of bio-based plastics are possible; these are available for both biodegradable and nonbiodegradable polymers (Table 7.2). The first is mechanical recycling; this process consists of well-known physical–mechanical steps such as the grinding, washing, separating. drying, regranulating, and

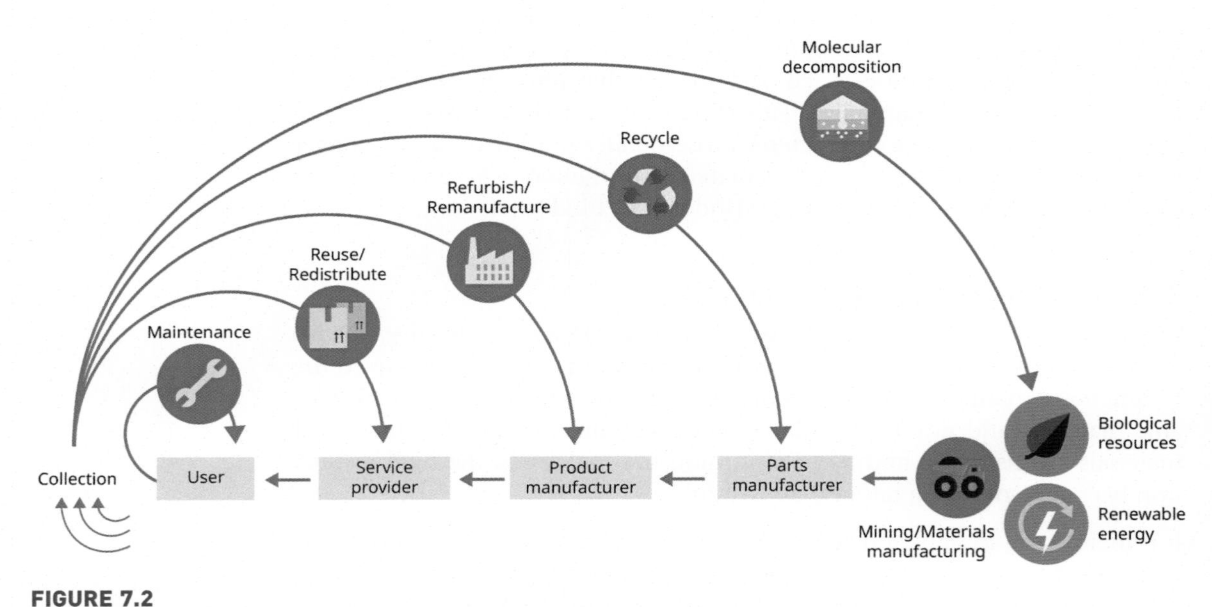

FIGURE 7.2
Adapted circular economy diagram giving an overview of available recovery pathways for products and materials. *Adapted from Ellen MacArthur Foundation.*

Table 7.2 Overview of recycling options for bio-based plastics (biodegradable and nonbiodegradable).

Recycling options	Description	Benefits and challenges
Mechanical recycling	The processing of waste plastics by physical means (grinding, shredding, melting) back to plastic products. To mechanically recycle postuser plastics waste, it has to be collected, sorted, separated, ground, washed, and reprocessed before it can be reused.	• Mainly for thermoplastic materials. • Prone to thermomechanical degradation which negatively affects the structure and morphology of the polymer and limits the number of recycling loops (La Mantia, 2004). • Results in prolonged carbon sequestration in products. • For biodegradable plastics it is important to determine whether they are sufficiently stable to be recycled in multiple cycles. • Some biodegradable plastics may cause contamination if they end up in the recycling stream of nonbiodegradable plastics. • Relatively low energy requirement.
Chemical recycling	Any process by which a polymer is chemically reduced to its original monomer form so that it can be processed (repolymerized) and remade into new plastic materials.	• Is currently being used for polyether, polyester, and polyamide (nylon). • Energy-intensive process. • The monomers can in some cases be repolymerized endlessly, giving them the qualities of virgin resin.

compounding of (thermoplastic) plastics. Usually, the resulting plastic properties are less well defined than those of virgin plastics. Second, for some plastics, especially condensation plastics, chemical recycling is possible. This is a depolymerization process that converts a plastic back into monomers, making it possible to repolymerize the polymer. This second option thus allows via a number of chemical processing steps the recovery of the pure polymer, but is more energy intensive than the first option.

7.3.3 Molecular decomposition

The molecular decomposition pathway looks at different ways that bio-based plastics can break down to a molecular level. The focus in this recovery pathway is the carbon cycle. Ideally, the decomposition of bio-based plastics is carbon neutral: when they break down, the carbon that was sequestered in the plastic becomes available again on a relatively short timescale (so, not hundreds of years).

There are five pathways (Table 7.3): plastics-to-fuel recycling, incineration with energy recovery, industrial biodegradation, anaerobic digestion, and unmanaged biodegradation. The first option, plastics-to-fuel recycling, is a thermal depolymerization process that results in biofuels and chemical feedstock, and which can be useful if the plastics are difficult to recycle in the recycling pathway. The resulting feedstock can be used for the production of new bio-based plastics. The second recovery pathway is incineration with energy recovery. Energy recovery aims to use the released energy obtained during the incineration (or "thermal" recycling) of plastics waste. These two routes are carbon neutral in the case of bio-based plastics, whereas they directly lead to an increased carbon footprint in the case of fossil-based plastics. The third and fourth recovery routes are only suitable for biodegradable plastics: industrial biodegradation and anaerobic digestion. Based on current knowledge and understanding, the fifth recovery pathway of "unmanaged biodegradation" should be avoided if possible, as it can lead to extremely slow and/or incomplete biodegradation.

Biodegradation is often considered a "regenerative strategy" for its ability to free up nutrients that feed the biocycle (EMF, 2013). From an economic perspective, however, the breakdown of a high-value material such as a bio-based plastic into carbon and water (or into other molecules such as methane) is essentially a form of value destruction. Designers therefore need to carefully consider how the products they develop will be used and recovered or discarded, and only specify biodegradable plastics if the application necessitates this. In all other cases, it may make more economic and environmental sense to design for product and material integrity as well as high-quality recovery, because this allows value to be retained (and captured) over a longer period.

Table 7.3 Overview of biodegradation options for biodegradable plastics.

Molecular decomposition options	Description	Benefits and challenges
Plastics-to-fuel recycling (or feedstock recycling)	Processes that turn plastic waste into bio-based oil or gas, which can be refined into biofuel, or converted into chemical feedstock such as bioethanol.	• Suitable for both biodegradable and nonbiodegradable bio-based plastics.
Incineration or thermal recycling (with energy recovery)	Combustion of plastic waste in incinerators. As plastics have a high energy content, the resulting heat from their combustion can be used directly (i.e., heating homes) or for the production of electricity.	• When bio-based plastics are incinerated with energy recovery, the energy produced is carbon neutral because CO_2 has first been sequestered into the bio-based plastic. • Incineration with or without energy recovery of biodegradable bio-based plastics has a similar climate effect as the incineration of nonbiodegradable bio-based plastics.
Industrial biodegradation (or industrial composting, or aerobic digestion)	This process takes place in a relatively short period of time and runs under specific process parameters with regard to humidity, temperature (exceeding 56°C), oxygen, etc. It is controlled by humans.	• Composting (under aerobic conditions) is CO_2 neutral. • Composting of bio-based plastics does in itself not produce compost (only CO_2 and water). • There can be cobenefits such as increasing the amount of food waste collected to be composted.
Anaerobic digestion (or anaerobic organic recycling)	The production of biogas by fermentation of biodegradable plastics. This process is controlled by humans.	• Not all biodegradable plastics can be digested through anaerobic digestion. • The process is CO_2 neutral if the biogas is subsequently used as fuel.
Unmanaged biodegradation (ambient conditions, can be aerobic or anaerobic)	The uncontrolled release of biodegradable plastics in the environment (soil, sea, space), where they may or may not break down or biodegrade in a reasonable timeframe under ambient conditions.	• Home composting of biodegradable bio-based plastics is often considered unmanaged biodegradation, because few people manage to actually control the heat and aerobic conditions in their compost bin to the extent needed for the composting of common biodegradable polymers. • Even perfectly biodegradable, bio-based polymers such as PLA and PBS can end up polluting oceans and soil if they biodegrade under uncontrolled conditions (Narancic et al., 2018).

7.3.4 Monstrous hybrids

Sidestepping the division between technocycle and biocycle of the butterfly diagram also eliminates another problematic concept that the circular economy inherited from Cradle to Cradle—monstrous hybrids. Monstrous hybrids are defined as mixtures of both technical and biological materials, neither of which can be salvaged after their current lives (McDonough, 2015), and which should therefore be avoided. But as most modern materials (including bio-based plastics) consist of composites or blends or are improved through additives, making a rigorous distinction between technical and biological materials is very difficult, if not impossible. On a product level this becomes even more complicated, as replacing all fossil-based plastics with bio-based plastics inherently will imply that plastic containing products in the techno-cycle will become monstrous hybrids. Thifs makes the concept of monstrous hybrids less useful for designers.

7.4 Designing with bio-based polymers in a circular economy: opportunities and challenges

There is an enormous diversity of biopolymers and as their production will inevitably be scaled up in the coming decades, the areas of application are expanding rapidly. Scientific developments indicate that bio-based equivalents of high-end engineering plastics, like Liquid Crystal Plastics and PEEK, will become available (Nakajima et al., 2017). Moreover, the production of bio-based plastics is no longer only motivated from a desire to substitute fossil resources, as we are seeing the development of bio-based polymers with improved or even completely new properties compared to fossil-based polymers. Examples are bio-based PEF and PTF that have excellent gas barrier properties and can substitute for PET (Andreeßen & Steinbüchel, 2018; Nakajima et al., 2017).

These developments create opportunities as well as challenges for industrial designers, which we will describe here. Among the many opportunities that bio-based materials offer, the one that stands out most is the fact that a completely new playing field for designers is opening up. Designers can begin to increase the demand for bio-based polymers by specifying "drop-in" bio-based polymers and bio/fossil blends (such as bio-PET). These polymers have similar properties to their fossil-based alternatives and their application is therefore without much risk. As a next step, designers can specify bio-based and biodegradable polymers for specific applications such as single-use packaging, hygiene applications, and agricultural products. This is more challenging because biodegradable polymers have different properties and because their recovery pathways can differ from traditional polymers. It requires designers to really understand and address the end-of-life options that are not open to nondegradable plastics, such as composting and anaerobic degradation (Narancic et al., 2018).

Taking a more forward-looking approach, there are opportunities for designers to work closely with material scientists to develop new kinds of bio-based polymers, for instance, the development of polymers with desirable characteristics that break down successfully in unmanaged conditions.

The collaboration between material scientists and designers will ensure that these new materials will not only meet functional specifications, but are esthetically and experientially pleasing as well, and will be applied in such a way as to maximize their potential (Barati & Karana, 2019).

In this chapter, we focused on bio-based plastics as these are most likely to help designers in substituting the current range of plastics used at a large scale in a variety of products. We would also like to briefly mention interesting developments in a complementary field of bio-based materials, which explores the use of living materials (Camere & Karana, 2018). Designers in this field showcase new ideas based on small-scale experimentation, for instance, the use of algae in building facades through building-integrated microalgae photobioreactors (Elrayies, 2018) or the use of fungi to create and grow products.

One of the major challenges that need to be overcome is the lack of knowledge among designers about the properties and possibilities of bio-based materials (and, in particular, bio-based plastics). Designers need to be brought up to speed on the quantitative physical properties of these materials, and not just on their experiential qualities, in order to enable collaboration with material scientists and industry. Bio-based plastics are not (yet) part of the curriculum in design schools, which may lead to a bias against bio-based plastics as having inferior properties. Next, the specification of a bio-based polymer always needs to be done on the basis of its characteristics *as well as* its recovery and end-of-life pathways. This too is a challenge—materials are usually specified based on their properties, without considering potential recovery pathways. If we are serious about achieving a circular and sustainable economy, this needs to change. Designers need to learn about the way bio-based plastics perform and degrade over time, how this affects their physical and experiential properties, and what recovery pathways are preferred given the application at hand. Finally, it should be stressed that the use of bio-based plastics is not necessarily a sustainable choice. For example, the impact of biodegradable plastics that are released into the environment in an uncontrolled way could be highly disruptive to natural ecosystems. Although bio-based plastics are made from renewable materials, those materials should come from sources that are replenished at a rate equal to or greater than the rate of depletion in order to be considered a sustainable choice. Overall, many sustainability and ethical challenges remain, and designers should make conscious decisions when specifying bio-based plastics.

7.5 Conclusion

We argue in this chapter that the world Ezio Manzini described in 1989, of "materials made to order," is about to be replaced by a world of "limits," where fossil-based plastics will no longer be available and where supply of alternative materials with negative sustainability impacts will be restricted. This may initially hamper design choice, but it will also open up a new and exciting world of bio-based materials waiting to be explored.

This chapter contributes to design practice and research in three ways. First, it positions the recovery pathways of bio-based plastics in a circular economy and in doing so creates a new conceptual model of the circular economy that combines "technocycle" and "biocycle" in one. The new model gives a clear overview of the different recovery options available for bio-based and biodegradable plastics. Second, we argue that the choice of a particular bio-based polymer always needs to be done on the basis of its properties *as well as* its recovery and end-of-life pathways, if we want to support a circular and sustainable economy. This is currently not a standard procedure in design. Finally, we argue that one of the main challenges we currently face is to educate designers about bio-based materials in all their complexity.

Through this chapter we hope to contribute to a renewed recognition of the materiality of design. The material basis of design was taken for granted for a long time, and designers may have grown somewhat complacent as a result. With the transition toward bio-based materials, the material basis is however back on the agenda and needs to be addressed urgently. It has never been more important that designers reengage with it in novel ways in order to help create a more sustainable future.

References

Andreeßen, C., & Steinbüchel, A. (2018). Recent developments in non-biodegradable biopolymers: Precursors, production processes, and future perspectives. *Applied Microbiology and Biotechnology, 103*, 143–157.

Azmah, J. S., Abdulla, R., Azhar, S. H. M., Marbawi, H., Gansau, J. A., & Ravindra, P. (2016). A review on third generation bioethanol feedstock. *Renewable and Sustainable Energy Reviews, 65*, 756–769.

Barati, B., & Karana, E. (2019). Affordances as materials potential: What design can do for materials development. *International Journal of Design, 13*(3), 105–123.

Camere, S., & Karana, E. (2018). Fabricating materials from living organisms: An emerging design practice. *Journal of Cleaner Production, 186*, 570–584.

European Commission (2018). *Report from the Commission to the European Parliament and the Council on the impact of the use of oxo-degradable plastic, including oxo-degradable plastic carrier bags, on the environment. COM(2018) 35 Final*, Brussels, 16.1.2018.

Elrayies, G. M. (2018). Microalgae: Prospects for greener future buildings. *Renewable and Sustainable Energy Reviews, 81*, 1175–1191.

Eurostat (March 4, 2019). Record recycling rates and use of recycled materials in the EU. *Newsrelease*, 39/2019.

Graedel, T. E., Harper, E. M., Nassar, N. T., & Reck, B. K. (2015). On the material basis of modern society. *PNAS, 112*(20), 6295–6300.

IRP (2019). *Global Resources Outlook; natural resources for the future we want*. A Report of the International Resource Panel. United Nations Environment Programme. Nairobi, Kenya.

Korhonen, J., Honkasalo, A., & Seppälä, J. (2018). Circular economy: The concept and its limitations. *Ecological Economics*, *143*, 37–46.

La Mantia, F. P. (2004). Polymer mechanical recycling: Downcycling or upcycling? *Progress in Rubber, Plastics and Recycling Technology*, *20*(1), 11–24.

Lambert, S., & Wagner, M. (2017). Environmental performance of bio-based and biodegradable plastics: The road ahead. *Chemical Society Reviews*, *46*, 6855.

Manzini, E. (1989). *The material of invention*. London: The Design Council.

McDonough, W. (2015). *Cradle to cradle product design challenge*. <https://mcdonough.com/cradle-to-cradle-product-design-challenge> Accessed 08.09.20.

Mülhaupt, R. (2013). Green polymer chemistry and bio-based plastics: Dreams and reality. *Macromolecular Chemistry and Physics*, *214*, 159–174.

Narancic, T., Verstichel, S., Chaganti, S. R., Morales-Gamez, L., Kenny, S. T., De Wilde, B., Padamati, R. B., & O'Connor, K. E. (2018). Biodegradable plastic blends create new possibilities for end-of-life management of plastics but they are not a panacea for plastic pollution. *Environmental Science Technology*, *52*, 10441–10452.

Nakajima, H., Dijkstra, P., & Loos, K. (2017). The recent developments in biobased polymers toward general and engineering applications: Polymers that are upgraded from biodegradable polymers, analogous to petroleum-derived polymers, and newly developed. *Polymers*, *9*, 523.

Sheldon, R. A. (2014). Green and sustainable manufacture of chemicals from biomass: State of the art. *Green Chemistry*, *16*, 950–963.

United Nations (2011). *Metal stocks and recycling rates*. Editor: International Resource Panel, Working Group on the Global Metal Flows, United Nations Environmental Programme.

CHAPTER 8

Biotextiles: making textiles in a context of climate and biodiversity emergency

Carole Collet
Living Systems Lab, Central Saint Martins UAL, London, United Kingdom

8.1 Introduction

With 10 years ahead of us to halve our greenhouse gas emissions as recommended by the International Panel for Climate Change (IPCC, 2018) and with the loss of one million species confirmed by the latest report from the Intergovernmental Science-Policy Platform for Biodiversity and Ecosystem Services (IPBES, 2019), every industry needs to explore and implement systemic change. The textile and fashion industry consists of a complex range of globally interconnected activities that all lead to detrimental and measurable impacts on land, water, biodiversity, and the atmosphere, alongside considerable human and social challenges. In the past 10 years, there have been unprecedented acknowledgments and efforts to better analyze and frame textile- and fashion-specific environmental challenges (The Sustainable Coalition, Ellen McArthur Foundation, Copenhagen Fashion Summit, The Sustainable Angle), with a growing number of global and local initiatives that support a transition to a more sustainable and circular industry (EU Textile Circular Economy, WRAP, Fashion Positive, Textile Exchange, Common Objectives) as well as funded support for disruptive innovation labs and startups (H&M Global Challenges, Fashion for Good, Kering Material Innovation Lab, LVMH Maison/0). The urgency of a global response to climate change is also more prominent than ever with activists such as Greta Thunberg and the Extinction Rebellion movement being front page news. Yet our growing world population and consumerist model keep increasing pressures on our natural resources: in three decades the global production of textile fibers has almost tripled (EEA, 2019) and the fashion industry has not demonstrated an ability to implement change fast enough to counterbalance its projected expansion (Lehmann et al., 2019, 2020). Sustainability is not enough anymore. Instead of minimizing our environmental impact using incremental improvements and cleaner technology, we need to rapidly progress to a restorative and regenerative mindset by

Materials Experience 2. DOI: https://doi.org/10.1016/B978-0-12-819244-3.00029-6

adopting a living systems approach so that what we design and make has a positive impact on our climate and biosphere (Birney, 2017; Pawlyn, 2019; Webster, 2017). This applies to the textile and fashion system which is on a current trajectory to use "26% of the carbon budget associated with a 2°C pathway" by 2050 (Ellen MacArthur Foundation, 2017, p. 21). We must envision and shape a new regenerative textile industry with disruptive design and fabrication strategies. And this is beginning to happen with the emergence of alternative propositions for textile materials and finishing processes arising from biodesign, biotechnology, and textile research.

This chapter examines key recent innovative bioalternative textile materials and processes and proposes an analysis of their potential contribution to lessening the environmental impact of the textile industry. With the emergence of biodesign and biotextiles, new questions arise. How can biodesign help reshape the textile industry? Can design-led biology research be scaled up to leverage circular principles for the bioeconomy? And how can biodesign help address textile-specific sustainable development goals? Section 8.2 situates textile material innovation in an environmental crisis context. Section 8.3 reviews the emergence of biodesign and its role in textile innovation, and Section 8.4 explores biotextile strategies mapped against living systems and biomimicry principles.

8.2 Textile material innovation in an environmental crisis context

8.2.1 From textile evolution to material revolution

Human evolution is deeply connected with the history of textile craft and production. Indeed, traces of dyed linen fibers have been found in the Dzudzuana cave in the republic of Georgia dating 32,000 YBP and anthropological research shows evidence that we started wearing cloth between 42,000 and 72,000 years ago (St. Clair, 2018). Crafting our textile materiality started with local knowledge but expanded with trade and global exchange. The silk road testifies to the influential development of textile fabrication across cultures with secrets of production fiercely guarded and/or stolen (St. Clair, 2018). Today, the textile industry includes small, cottage local industries as well as large global multinational companies with complex tiered supply chains, but it still relies on material knowledge and processes mastered long ago, such as dyeing, spinning, weaving, and knitting.

When it comes to sourcing raw materials for textiles, we can categorize textile production as follows (see Fig. 8.1):

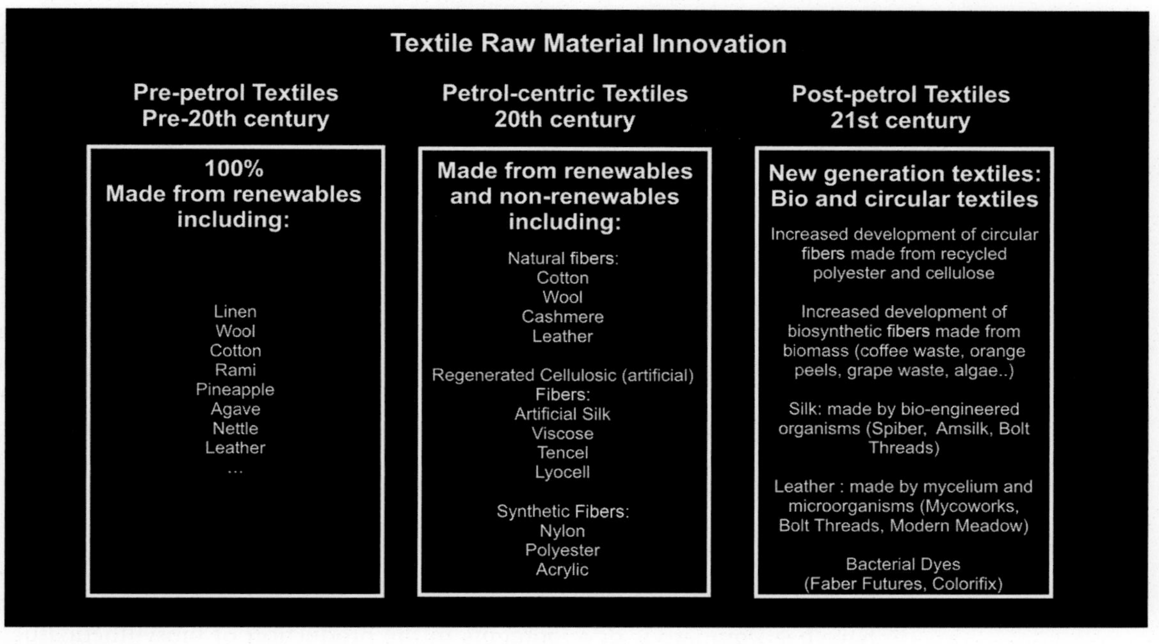

FIGURE 8.1
Mapping key textile raw material innovation pre- and post-petrol economies (Collet, 2019). Data gathered via the literature review of textile material innovation, biotechnology, and biodesign.

- a pre-petrol textile era where materials are made from natural renewable sources;
- a petrol-centric era where synthetic fibers dominate the production of textiles, and where petrochemistry is also heavily used in the production of natural fibers such as cotton in the form of synthetic pesticides and fertilizers;
- a post-petrol era where recycled, circular, and bioinformed new materials and processes are fast emerging as alternatives to unsustainable conventional textiles.

Until the end of the 19th century, textiles had been made from natural materials. They were either plant-based such as linen, cotton, rami, bark cloth, pineapple, agave, and nettle; or animal- and protein-based for the production of wool, cashmere, silk, and leather. Material innovation laid in the skills and ability to know how to transform harvested or foraged materials into functional textile fibers. How do we process a flax plant to extract its fiber? How do we cultivate silkworms? However, in 1855, a major invention took place with the development of artificial silk or rayon (made from chemically processed plant cellulose), which was first commercially produced in 1910 (Shaikh et al., 2012). From the cave of Dzudzuana up until 1855, plant- and animal-based materials were transformed into fibers by mechanical or handmade processes, not by chemistry. The invention of artificial silk broke thousand years of tradition in terms of harvesting and sourcing material for the production of fibers and marked the beginning of a quest for a different kind of raw material sourcing. This was to come with the invention of synthetic fibers starting with Nylon in 1935 (Raber, 2010) and Polyester in 1941 (Sterlacci & Arbuckle, 2017). Since its invention, Polyester has radically disrupted the entire textile production system. Made from nonrenewable crude oil it is much less expensive to produce than natural fibers, and as such has facilitated the development of fast fashion. Today, 70 million barrels of oil are used every year to produce polyester clothing and over 8000 different chemicals are used to turn raw material into textiles with 25% of the world's chemicals used for textile production (Common Objectives, 2018a). Within a few decades, the millennia-old textile industry has shifted from a reliance on natural renewable fiber production to a dependence on nonrenewable crude oil.

By the end of the 20th century and the beginning of the 21st century, there have been numerous developments in the production of biosynthetic fibers. Biosynthetics are polymers made from sugar, starches, and lipids from renewable sources such as corn, sugar, cane, beets, and plant oils; and although not new, they are becoming a viable alternative to oil-based synthetic fibers (Textile Exchange, 2018c). We are seeing increasing development of new types of biosynthetics made from waste biomass such as coffee, orange, or grape waste. But where radical textile material innovation lies is in the development of a new breed of fibers derived from research in synthetic biology. Made with living organisms and relying on the inherent properties of biological organisms to biofabricate at ambient temperature with no toxic by-products, these new biotextiles have galvanized public interest. Key collaborations across biotech companies and fashion/sportswear brands have transformed unrivaled hi-tech material innovation into fashionable and desirable products. For instance, biotech firm Spiber collaborated with the North Face to launch the Moon Parka in 2015. It is made with Brewed Protein silk (Rhodes, 2015) and was shortly followed in 2016 by the launch of Adidas first shoe made with Biosteel by Amsilk. In 2016, Bolt Threads biofabricated the first tie made with silk grown by yeast, and later partnered with Stella McCartney to design a dress made with their bioengineered Microsilk for the milestone exhibition: Items: Is Fashion Modern at the Museum of Modern Art. The partnership continued when Adidas and Stella MacCartney launched the first biodegradable Microsilk tennis dress in 2019. These radical new biotextiles will be discussed in Section 8.2.

8.2.2 Making textiles for a growing world population

The current environmental impact of the textile and fashion industry has become more widely recognized by policy-makers, consumers, and manufacturers, yet, as early as the 1800s, "evidence of public disquiet about pollution from the textile industry can be found in newspapers" (Erhman, 2018, p. 45). Business values and economics prevail above the protection of nature. Inscribed within a culture of endless material growth, where "Nature is accorded an almost infinite capacity to take care of various wastes and pollutants and render them harmless" (Wijkman & Rockström, 2012, p. 4), the production of textiles whether natural, artificial, or synthetic has led to fundamental and critical environmental issues. "Textiles are the fourth highest pressure category for the use of primary raw materials and water, after food, housing and transport, and fifth for greenhouse gas emissions" (EEA, 2019, p. 10).

There are two key factors to consider in terms of textile raw material provenance. First, the environmental impact directly linked to the production process of the raw material. For instance, cotton that accounts for 25% of global fiber consumption is highly damaging in terms of biodiversity, land toxicity, and water scarcity when produced with conventional petrochemical intensive agricultural methods. In addition to a large amount of water, it uses high levels of pesticides, fertilizers, and chemicals that affect human health and pollute local ecosystems (Shepherd, 2019). Polyester represents over half of the entire fiber market and over 75% of all synthetic fibers. This equals a sevenfold growth since 30 years ago (Textile Exchange, 2018c, p. 9). Made from crude oil, a limited nonrenewable supply, it has a high carbon footprint: emissions of CO_2 for polyester in clothing, at 282 billion kg in 2015—are nearly three times higher than those for cotton, at 98 billion kg (Cobbing & Vicaire, 2016, p. 4). Although recycled polyester is an alternative to using virgin oil, a critical issue linked to synthetic fibers is their release of microfibers into water streams during the washing cycle. The accumulated amount of microfiber entering the ocean between 2015 and 2050 would exceed 22 million tons—about two-thirds of the plastic-based fibers used to produce garments annually (Ellen McArthur Foundation, 2017, p. 39). Second, we need to take into account the growing quantity of the global material flow related to population and consumption growth patterns: global fiber production saw a 10-fold increase from 1950 to 2017 from <10 million mt to over 100 million mt (Textile Exchange, 2018a, p. 6). Fig. 8.2 maps raw textile material innovation together with world population which is predicted to reach 9.7 billion people by the year 2050 (United Nations Worldometer, 2020).

We can see that at the time of the first commercialization of artificial silk in 1910, the world counted 1.7 billion people and less than 2.5 billion when polyester came to market in 1941. Sixty years later, the population has reached

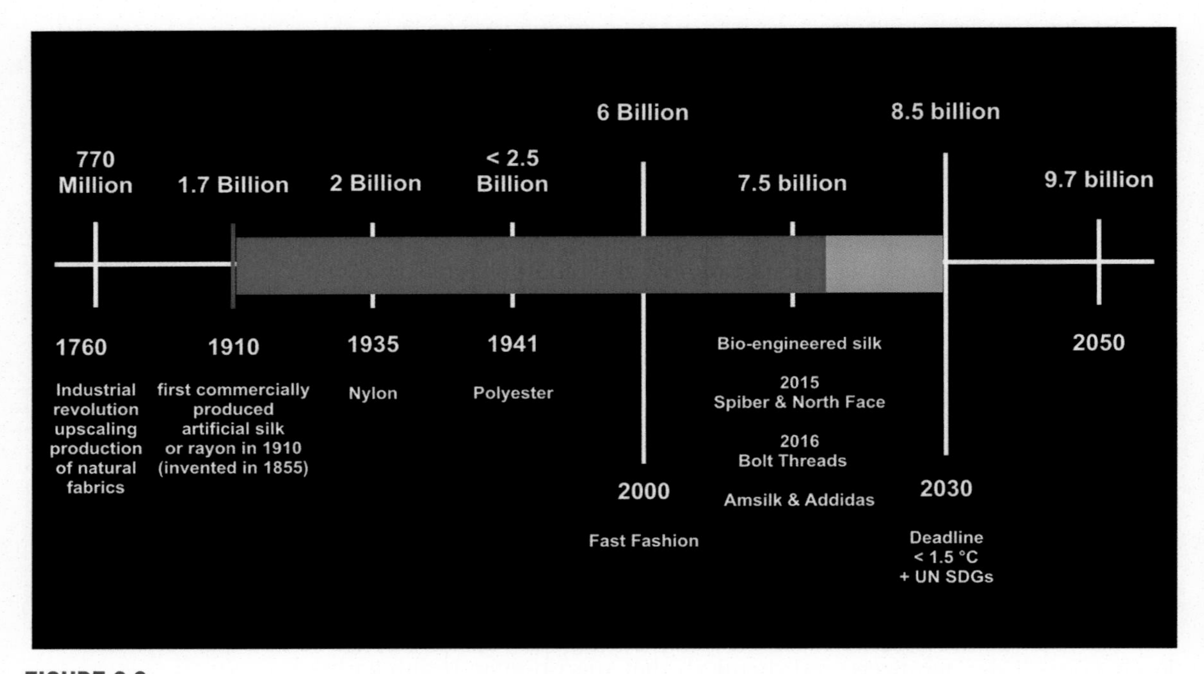

FIGURE 8.2
Mapping key textile material innovation with world population growth (Collet, 2019). Data gathered from: United Nations, from 1950 to current year: elaboration of data by United Nations, Department of Economic and Social Affairs, Population Division. World Population Prospects: The 2019 Revision; from Worldometers (http://www.worldometers.info) and literature review of textile material sourcing.

6 billion. In 2015–16, when the first bioengineered silk made by microorganisms is launched (Spiber, Amsilk, Bolt Threads), the world counted 7.5 billion humans (Worldometer, 2020). By 2030, when we should have halved our greenhouse gas emission to stay within the 1.5°C temperature increase (IPCC, 2018) we will be an extra 1 billion compared to 2015. So how do we generate textiles and clothing for another 1 billion people while transitioning to a post-carbon economy? Nothing short of a material revolution is needed and it has begun to materialize with advances in biological sciences paralleled with the emergence of biodesign. Section 8.3 will develop how biodesign and biotextile alternatives are beginning to disrupt the textile industry.

8.3 Textile innovation in the context of the emergence of biodesign

While cotton and polyester remain forecasted as the leading fibers for the foreseeable future (Ellen MacArthur Foundation, 2017; Grose, 2015;

Textile Exchange, 2018c), environmental pressures have become a catalyst for the development of alternative textiles. In a shifting context where a transition to the circular economy promotes recycling, upcycling, and closed-loop systems, there is an increase in the use of recycled raw materials, cotton and polyester included (Textile Exchange, 2018a). However, this section will focus specifically on the more radical recent textile material innovations that exist at the crossing of design and biology and will ask: how do we design for a post-carbon textile bioeconomy using the intrinsic ecological values of biological systems (cyclic, solar, local) to develop a regenerative blueprint for textiles? For the purpose of this chapter, we will examine biodesign and more specifically biotextile innovation within a 10-year timeframe.

2010 marked a profound step in human history, when US scientist Craig Venter and his team claimed the creation of the first synthetic organism on Earth, designed on a computer (Pennisi, 2010). This breakthrough marked a time when biology became a powerful technology, one that we aim at controlling to generate new biologically driven microfactories that can be programmed, localized, and customized (Collet, 2015, p. 197). In parallel to scientific advances, the field of biodesign began to emerge. Informed by strategies found in nature, biodesign goes beyond referencing nature and relies on active collaboration with biological organisms to propose new ways of making (Myers, 2012). Biodesign was celebrated in the inaugural exhibition *Alive, New Design Frontiers* at the EDF Foundation curated to reevaluate the process of design when confronted with the "living" (Collet, 2013, p. 6). By understanding that "design can be a means to unlock new world of knowledge but also an efficacious tool for managing current technological transformations and transgression" (Doll et al., 2016, p. 1) we can explore the potential of biodesign to offer novel perspectives on what change could look like, for ourselves and other living things (Ginsberg & Chieza, 2018).

Inscribed within a discourse of biomimetics, the field of biodesign is recent enough that its definition evolves as it expands and the "phenomenon of growing design is still scarcely understood" (Camere & Karana, 2017, p. 101). Here, we understand biodesign "as a means to incorporate the inherent life-conducive principles of biological living systems into design processes—to transition into a more holistic, sustainable future" (Collet, 2019). While multifarious terms are used, such as bioinformed design, biointegrated design, growing design, biomimetic design, each with their own interpretation of a design activity hybridized with biological principles and/or living systems, there are two key streams of innovation. One is led by designers, and the other is led by scientific research and biotechnology. Both have adopted the terminology of "biodesign" to manifest biointegrated processes. Biodesigners have been particularly active in the field of textiles and have collaborated with a range of living systems to reinvent our future textile materiality with the help of organisms such as bacteria (Lee, Chieza, Keane) and mycelium (MycoTEX, Montalti, Collet). Others have incorporated inert biomass into the development of new materials such as algae (Algiknit, Vegea, Orangefiber). Synthetic biologists, on the other hand, have pushed the limits of biology to create plastic with methane eating bacteria (Mango Materials), or reprogram yeast to express silk proteins (Bolt Threads, Amsilk, Spiber), or to scale up biological processes that can radically alter textile pollution resulting from the dying process (Colorifix, Pili, Tinctorium).

As the field of biodesign grows, we can begin to develop a new interpretation of material innovation by mapping different design and fabrication strategies. Curatorial research developed for the exhibition discussed above *Alive, New Design Frontiers* in 2013 referred to five key design strategies to negotiate design relationships with the natural world and to locate designers operating "within a sliding scale of a natural nature and a new programmable nature in the quest for innovative ecological models" (Collet, 2013, p. 6). This initial framework was devised to map the practice review of the very nascent field of biodesign when developing the proposal for the exhibition in 2011. *The plagiarists, the new artisans, the bio hackers, the new alchemists*, and *the agents provocateurs* were the titles used in the exhibition to communicate biodesign strategies to a broader public. They correspond to five different approaches: nature as a model, a coworker, as well as reprogrammed, hybridized, or conceptualized nature. The choice was made to include a "conceptualized nature" section as many designers, short of accessing actual biology laboratories and tools, were exploring speculative concepts to develop an imaginary of biotechnology and propose an ecological bioinformed future. Three years later the framework was simplified with three key strategies (Fig. 8.3), excluding conceptualized and fictional nature as more designers began to actively engage with living systems and developed a "DIY-Material practice" (Rognoli et al., 2015). This testifies to the rapidity of the development in the field with pioneers Suzanne Lee, Phil Ross, and Maurizio Montalti at its helm.

As the field continues to expand, so does its taxonomy. Another key and more recent model proposes four interconnected approaches to biodesign: Growing Design, Augmented Biology, Biodesign Fiction, and Digital Biofabrication (Camere & Karana, 2018). This model provides a more agile intersected framework centered around the practice of biodesign and the possibilities to merge different approaches together.

8.4 Strategies for designing textiles with living systems

When it comes to textiles, the conventional and classical taxonomy refers to three main categories of materials based on their nature and origin: natural

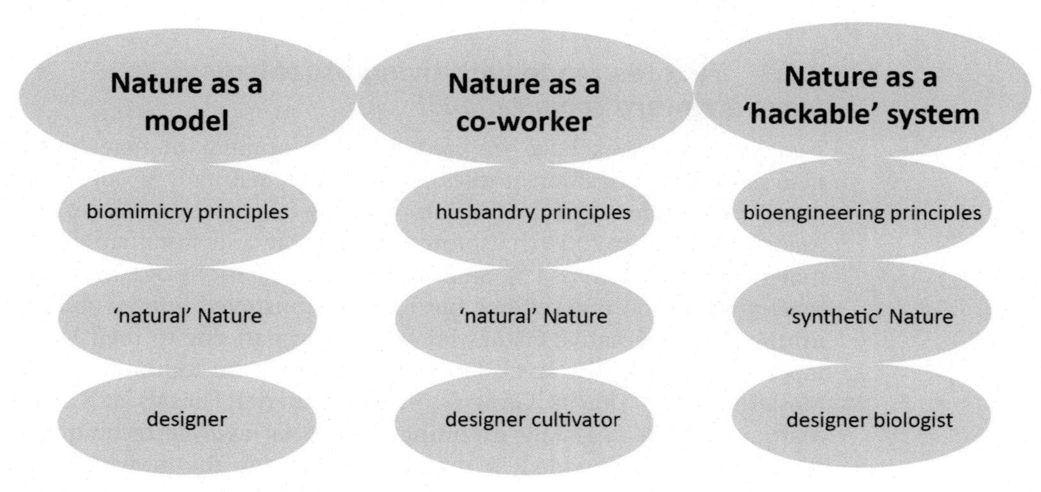

FIGURE 8.3
From Biomimicry to Biofacture. © Carole Collet, Design & Living Systems Lab, Central Saint Martins UAL, 2016.

fibers, (agriculture), regenerated artificial fibers (manufacture), and synthetic fibers (manufacture). In 2015 a revisited version informed by the latest development in synthetic biology and the UK roadmap for synthetic biology was published to include biofiber and biofabric categories for biofactured material systems (Collet, 2015, p. 194) and situates textiles in a context where there is a disconnect "between the ways in which clothing is made and the ways in which it *might* be made" (Antonelli, 2017, p. 9). There are many other ways to map textile materials; here we are concerned with an examination of biotextile materials in relation to environmental issues. How do we examine the sustainable value of biodesign and biotextile propositions in relation to material innovation and origins? Below is a list of key recent biomaterial innovations examined via the lens of principles of natural living systems (Benyus 1997; St. Pierre, 2015, p. 40). Benuys advocates the adoption of nature's law to address critical human-made ecological challenges. "Nature runs on sunlight. Nature uses only the energy it needs. Nature fits form to function. Nature recycles everything. Nature rewards cooperation. Nature banks on biodiversity. Nature demands local expertise. Nature curbs excess from within. Nature taps the power of limits." (Benyus, 1997, p. 7). The examples below do not represent an exhaustive list of biotextile innovations but are referenced as a means to develop a living system critique of biotextile practices as well as to identify key research gaps. By adapting

Benyus's principles into design questions, we can begin to frame the emergence of biotextiles in relation to biomimicry principles.

8.4.1 Textiles made with living organisms: *(does it reward cooperation?)*

Two key organisms have inspired the development of new biofabricated material systems' fashion textiles. Bacteria that naturally produced cellulose were first introduced to the fashion world by Suzanne Lee via the Biocouture Atelier founded in 2003 to explore compostable and sustainable biofactured materials (Lee, 2013, p. 19). Since then, many iterations of bacterial cellulose materials have been developed but have demonstrated limited ability to scale up to viable alternative textiles for fashion, in part due to their lack of resistance to the humidity level. Jen Keane revisited the technique for her final Master project "This Is Grown" on MA Material Futures at Central Saint Martins UAL in 2018. By inventing a *microbial weaving* technique where a fiber matrix is embedded in the cellulose during the growing phase. This allows for the production of stronger biodegradable materials that are also shaped during their growth: "nature doesn't make its materials in sheets, dye, cut and assemble. It produces just what it is needed as it is needed" (Keane, 2019, p. 4). "This Is Grown" has been further developed with two collaborations. A residency at Bolt Threads in 2019 helped to further refine and develop microbial weaving techniques and includes using Bolt Threads Microsilk in the woven matrix (Figs. 8.4 and 8.5).

Another collaboration was established with the synthetic biologist Marcus Walker, PhD candidate in the Tom Ellis Lab at Imperial College London in 2019 with the support of Mills Fabrica and Hong Kong Innospace. Walker "has used genetic engineering techniques to develop a self-dyeing bacterium that produces both cellulose and melanin, a natural pigment found in squid ink, hair, and skin. Employing this bacterium in Keane's microbial weaving process, together they have grown the first sneaker-upper woven and dyed by a single genetically modified organism" (Keane, 2019, p. 7) (Figs. 8.6 and 8.7).

Jen Keane's work embodies two different types of symbiotic cooperation. First, in the material itself, as bacterial cellulose is produced by an ecosystem of bacteria and yeast that depend on one another to thrive. Second, in the collaboration between synthetic biologists who bring an expert understanding of biological systems and the designer whose knowledge of the material and design world can guide and respond to scientific experiments. The intersection of different sets of expertise provides a new type of cooperation that can help discover new approaches to making textiles.

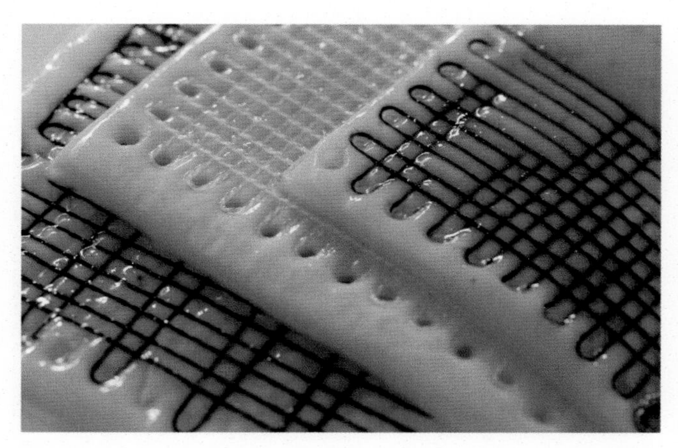

FIGURE 8.4
This is Grown. © Jen Keane. Microbial woven samples produced by Jen Keane during her residency at Bolt Threads 2019.

FIGURE 8.5
This is Grown. © Jen Keane. Microbial woven samples produced by Jen Keane during her residency at Bolt Threads 2019. Details of Microsilk (yellow fiber) embedded in bacterial cellulose.

Parallel to bacterial cellulose developments, there are many successful examples of artists, designers, and biotechnologists exploring the potential of living mycelium to grow materials. Pioneered by artist Phil Ross who integrated mycelium in his practice in the 1990s before to establish Mycoworks in the United States. In the past decade, the majority of mycelium research has focused on the design and production of solid materials for product,

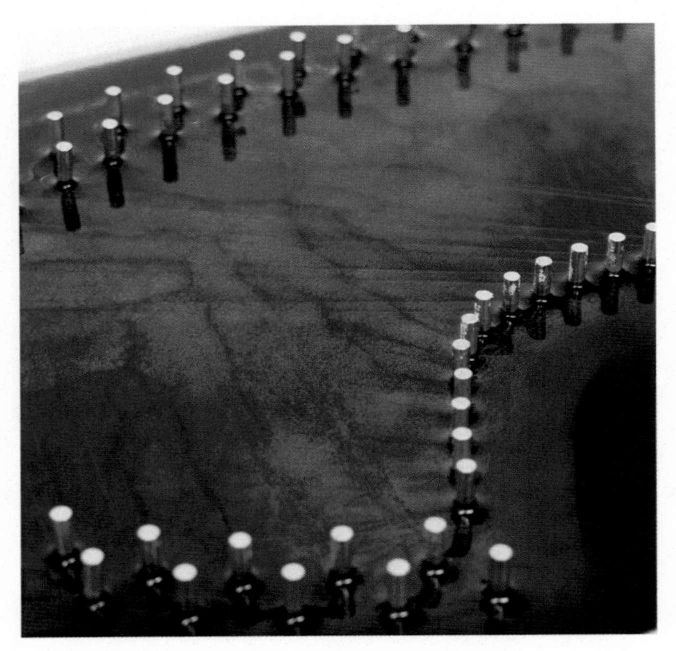

FIGURE 8.6
This is GMO. Jen Keane in collaboration with Marcus Walker. Detail of the liquid microbial culture. Photography: Jen Keane.

packaging, and architectural design proposals (Ecovative, Ross, Montalti, Mogu, Klarenbeek, Benjamin). In terms of textiles, Aniela Hoitink designed the first dress made with mycelium material and was awarded the H&M Global challenge with MycoTEX in 2018 (Global Change Award). Ross launched the first mycelium-based leather material with Mycoworks in 2016, shortly followed by the introduction of Mylo made by Bolt Threads. Ecovative and Mogu have also developed foam-like materials. The latest leather, Reishi, was launched for New York fashion week by Mycoworks in February 2020. As a vegan alternative to animal-based leather, mycelium leather is one of the most promising new biotextiles in terms of sustainability. It can grow on waste biomass in a matter of days compared to months for the production of cow hides which are detrimental to our biodiversity (deforestation and land use to produce food stock for cattle) and have a high environmental cost (Common Objectives, 2018b). Cooperation with mycelium has proven to be an effective means to develop new biocircular textile materials, which can be upscaled from laboratory research to industry standards and propose concrete, viable alternatives for sustainable textile fabrication processes.

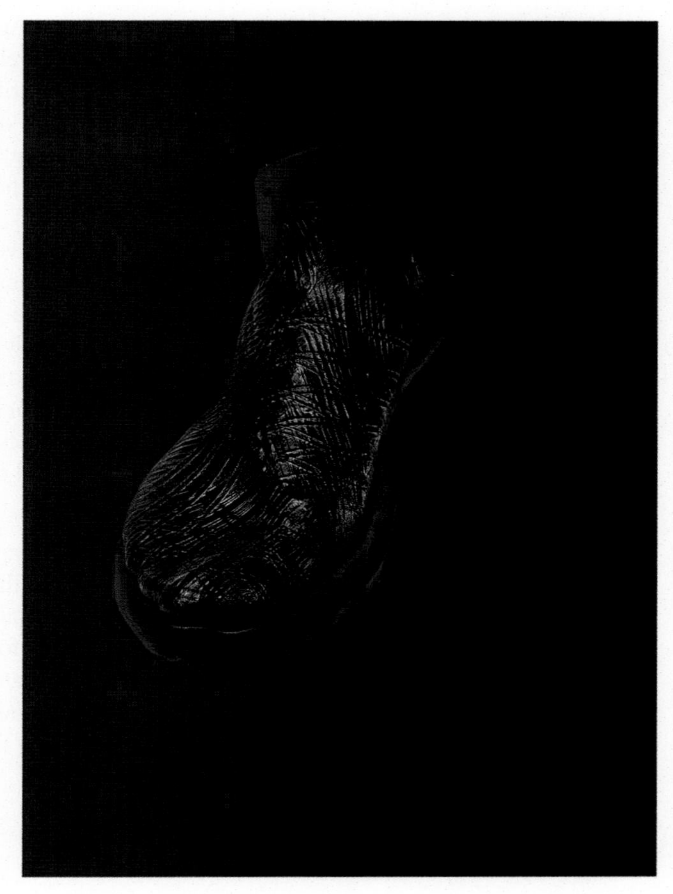

FIGURE 8.7
This is GMO. Jen Keane in collaboration with Marcus Walker. Photography: Ed Tritton.

8.4.2 Textiles patterned with living systems: *(does it reward cooperation?)*

Textile finishing processes that include dying, printing, and coating currently rely on petrochemical processes which left untreated are toxic for our water effluents. Natural dyes do not offer a realistic alternative to synthetic dyes due to the sheer volume required at global scale. Traditional textile dye recipes require the use of mordants to help fix the dye onto fibers and become color fast. These mordants include heavy metals, which are toxic for the environment. A huge amount of plants would also be needed and compete for land currently used to grow food. So natural dyes are not a viable alternative, but cooperation with living microorganisms can help us reinvent new processes to pattern our clothes. Designer Natsai Audrey Chieza, founder of studio Faber Futures, has been promoting the development of microbial dye since 2011. In collaboration with Professor John Ward, University College London she developed a series of recipes to dye silk with bacteria. *Project Coelicolor* culminated at Gingko Bioworks where Chieza initiated a residency to explore how to transition from producing petri dish size samples to a length of microbial dyed fabrics (Faber Futures). Her influential work embodies the quest for more sustainable and alternative ecological futures as she was awarded the Index Award 2019 for the chemical-free and water-saving dye process she developed as a designer (The Index Project, 2019). With the same ambition, several startups are now driving disruptive change for the textile finishing sector (Pili, Tinctorium, Colorifix). Using synthetic biology tools Colorifix programs microorganisms to both grow and fix a range of colors onto either natural or synthetic fabrics (Fig. 8.8) using "no additional specialist equipment or toxic chemicals and one-tenth of the water of standard processes" (Colorifix, 2020).

Mycelium Textiles (Collet 2016–2019) explores another approach for textile patterning with living systems and investigates the possibilities of combining the dynamic active properties of mycelium with traditional textile know-how to develop sustainable textile embellishment techniques for a post-petrol textile industry. As discussed above, research into mycelium materials is prolific, but its use as a biopatterning process for textiles is an underresearched area. Results include the production of the self-patterning mycelium rubber, the production

FIGURE 8.8
Textile samples dyed with engineered microorganisms. © Colorifix.

FIGURE 8.9
Tie-Grow on cotton. © Carole Collet, Design & Living Systems Lab, Central Saint Martins UAL.

of a permanent pleating process at ambient temperature for cotton, a "tie-grow" mycelium on cotton (Fig. 8.9), the development of an embroidery-like patterning effect on cotton, and the use of mycelium as a binding agent to assemble mixed textile materials to create a lace-like textile technique.

Cooperating and collaborating with living systems to develop alternative textile finishing processes is still a niche area but a crucial one. Current finishing processes that include dying, printing, and coating require a large amount of water and the fashion industry with a water usage set to double by 2030 is recognized as the third largest user of water globally (Common Objectives, 2018c). Biocompatible, low energy, low water usage, and biodegradable processes informed by laws of nature can make a real impact and further research is needed in this area.

8.4.3 Textile fibers and materials made by reprogrammed organisms: *(does it use only the energy it needs?)*

As discussed in Section 8.2.2, synthetic biology research disrupted textile material innovation with the introduction of a new type of fibers and materials brewed by proteins and grown in a lab. Spiber, Amsilk, Bolt Threads, and Modern Meadow opened the door to a new programmable textile materiality, sourcing DNA codes, and cellular protocols evolved over 3.8 billion years to produce new kinds of natural fibers using genetically engineered microorganisms as microfactories. If the output material is "natural," the "microfactory" is genetically engineered. Beyond the scientific exploit, we need to examine the relevance of this radical innovation. One key tenet of the laws of nature is that the natural system uses less energy than human-made technological systems. "In technology about 75% of problems are solved by manipulating energy, whereas the equivalent figure for biological systems is 5%" (Vincent, 2014, p. 241). So, relying on biologically active microorganisms to grow textile materials is in theory a more energy-efficient system. However, the production of these new biofibers and materials has not yet been scaled up to large-scale manufacturing and their full life cycle analysis is incomplete. They require sugars to grow, so the production cycle of these sugars needs to be carbon negative if we are to offer a genuine alternative to shape a future regenerative textile and fashion industry and it is still too early to balance the pros and cons of genetically engineered brewed fibers compared to conventional textile fibers in terms of full environmental impact. This is also an area where further research is needed.

8.4.4 Textiles made by and for emission capture: *(does it curb excess from within?)*

Here, we will discuss a series of cutting-edge biotextile innovations that rely on biocapture to biofabricate materials. Mango Materials, for instance, uses a bacterium that feeds on waste methane gas to produce a biodegradable biopolyester. Wearpure has developed a biopolymer suitable for three-dimensional printing textiles, which can mineralize primary greenhouse gases (CO_2 and NO*x*) and reduce volatile organic compounds. DyeCoo has patented a water-free dyeing process using reclaimed CO_2 as the dyeing medium in a closed-loop process. The startup Post Carbon Lab is working on developing a Photosynthesis Coating that transforms textiles into carbon-capture surfaces. Designer Roya Aghighi collaborated with the University of British Colombia and Emily Carr University, to create Biogarmentry, a textile research project that uses *Chlamydomonas reinhardtii*, a type of single-cell green algae, which when activated by sunlight can photosynthesize and turn CO_2 into oxygen. While this is a rich context for research, some of these biotextiles are still at the proof-of-concept stage. We are far away from a widespread use of carbon-capture textile materials, which by default would only make a real difference to the fashion carbon footprint when scaled up. However, exploring nature-inspired strategies to use textiles as carbon sink and carbon-capture material is a relevant path. The question is when will these biotextile innovations be ready for large-scale adoption?

8.4.5 Textiles made with abundant material flows: *(does it support biodiversity?)*

We will conclude this section by exploring new biomaterials made from some of the most abundant materials found in nature. The most common chain molecules in the world are cellulose and chitin (Fratzl, 2016, p. 173). Elissa Brunato works with cellulose and has created the first bio-iridescent sequins that produce shimmering structural colors for a graduate project on MA Material Futures at Central Saint Martins UAL in 2019. Collaborating with material scientists Hjalmar Granberg and Tiffany Abitbol from the RISE Research Institutes of Sweden, her key objective was to develop an alternative to plastic sequins commonly used in the fashion industry. Sequins are detrimental to ocean life and contribute to microplastic pollution. The bio-iridescent sequins offer a biodegradable and compostable alternative (Fig. 8.10).

Algae also occur in large quantities on Earth, even more so with algal blooms triggered by the use of fertilizers in agriculture. Creating materials derived from algae offer a new perspective for fashion. Algiknit, for instance, integrates biotechnology and design to develop algae-derived yarns that are biodegradable and compostable while Algix makes shoe insole out of algae.

By capitalizing on the use of naturally abundant materials instead of investing in oil-based technologies, these examples showcase that we can shift the textile and fashion system and develop biodiversity-positive new material propositions. Here, biodiversity is enhanced by two strategies: the replacement of plastic with biodegradable alternatives and the use of excess living materials such as algal blooms (the result of intensive agricultural practice) to replace synthetic polymers.

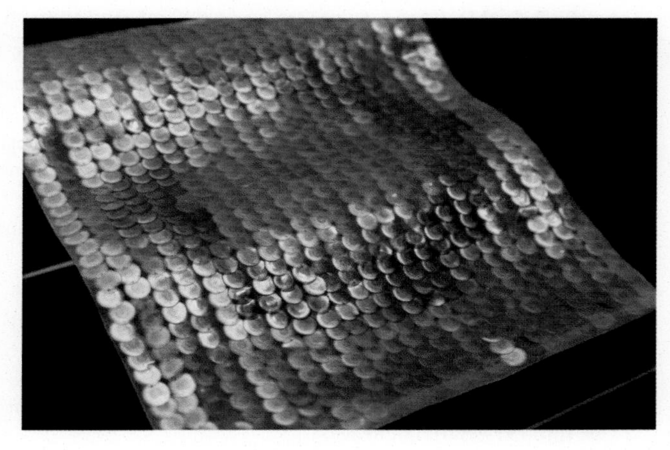

FIGURE 8.10

A hand-embroidered sample using bio-tridescent sequins, showing the sequins natural shimmering colors. © Elissa Brunato.

8.5 Conclusion

This chapter posits that we are at a crossroad between an oil-centric and a post-petrol textile industry. By first asserting the damaging effects of the textile and fashion industry on land, water, biodiversity, and climate change, we have also established that despite valid positive initiatives and new frameworks, the industry is not changing fast enough to meet the recommendations of the International Panel for Climate Change in terms of carbon emission. Incremental change, cleaner, and more efficient technologies are not enough. Looking more closely at textile material innovation mapped against human population, we recognized the urgency to explore radical new systems that can disrupt the current textile production paradigm. Recent scientific advances, especially in the field of synthetic biology and the eagerness of a new generation of designers to search for and create new materials, have contributed to the emergence of biodesign. A fast-evolving discipline, biodesign research has been prolific in developing biotextile material innovations in the past 10 years, often in response to environmental concerns. While it is still too early to fully analyze their environmental benefits, we can examine their credibility against living systems principles derived from biomimicry. As such this chapter provides a new insight into biotextile strategies driven not by their material origin, but by their ability to reward cooperation, their actions onto the environment (such as carbon capture), or their ability to enhance and support biodiversity.

References

Antonelli, P. (2017). *Items: Is fashion modern?* New York: The Museum of Modern Art.

Benyus, J. (1997). *Biomimicry. Innovation inspired by nature*. New York: William Morrow.

Birney, A. (2017). *Cultivating system change. A practitioner's companion*. Abingdon: Greenleaf Publishing.

Camere, S., & Karana, E. (2017). Growing materials for product design. In *Proceedings of the international conference of the Design Research Society Special Interest Group on Experiential Knowledge (EKSIG)*, Delft University of Technology, 19–20 June.

Camere, S., & Karana, E. (2018). Fabricating materials from living organisms: An emerging design practice. *Journal of Cleaner Production, 186*, 570–584.

Cobbing, M., & Vicaire, Y (2016). *Timeout for fast fashion*. Greenpeace.

Collet, C. (2019). *MA Biodesign course summary*. <https://www.arts.ac.uk/subjects/textiles-and-materials/postgraduate/ma-biodesign-csm> Accessed 29.03.20.

Collet, C. (2015). The new synthetics: Could synthetic biology lead to a sustainable textile manufacturing? In K. Fletcher, & M. Tham (Eds.), *Routledge handbook of sustainability and fashion* (pp. 191–200). Routledge.

Collet, C. (Ed.). (2013). *Alive, new design frontiers*. EDF Foundation. <https://www.thisisalive.com> Accessed 04.03.20.

Colorifix (2020). <https://colorifix.com> Accessed 20.03.20.

Common Objectives. *The issues: Chemicals.* (2018a). <https://www.commonobjective.co/article/the-issues-chemicals> Accessed 04.03.20.

Common Objectives. *Fibre briefing: Leather.* (2018b). <https://www.commonobjective.co/article/fibre-briefing-leather> Accessed 04.03.20

Common Objectives. *The issues: Water* (2018c). <https://www.commonobjective.co/article/the-issues-water> Accessed 04.03.20.

Doll, N., Bredekamp, H., & Schaffner, W. (Eds.) (2016). *+ ultra. knowledge & gestaltung*. Catalogue exhibition. Seemann.

EEA (2019). *Textiles in Europe's circular economy.* <https://www.eea.europa.eu/themes/waste/resource-efficiency/textiles-in-europe-s-circular-economy> Accessed 11.03.20.

Ellen MacArthur Foundation (2017). *A new textiles economy: Redesigning fashion's future.* <http://www.ellenmacarthurfoundation.org/publications> Accessed 11.03.20.

Erhman, E. (2018). *Fashioned from nature*. London: V&A Publishing.

Fratzl, P. (2016). The bioinspired design of material. In N. Doll, H. Bredekamp, & W. Schaffner (Eds.), *+ ultra. knowledge & gestaltung*. Catalogue exhibition. Seemann.

Ginsberg, A. D., & Chieza, N. (2018). Editorial: Other biological futures. *Journal of Design and Science*. Available from https://doi.org/10.21428/566868b5, Accessed 04.03.20.

Grose, L. (2015). Fashion as material. In K. Fletcher, & M. Tham (Eds.), *Routledge handbook of sustainability and fashion* (pp. 223–233). Routledge.

IPBES Intergovernmental Science-Policy Platform on Biodiversity and Ecosystem Services. *Summary for policymakers of the global assessment report on biodiversity and ecosystem services*. (2019). <https://doi.org/10.5281/zenodo.3553579> Accessed 11.03.20.

IPCC Intergovernmental Panel on Climate Change. *Special report: Global warming of 1.5°C—Summary for policymakers.* (2018). <https://www.ipcc.ch/sr15> Accessed 11.03.20.

Keane, J. (2019). *This is grown*. Press release obtained in email correspondence with author.

Lee, S. (2013). In C. Collet (Ed.), *Alive, new design frontiers*. Exhibition catalogue. EDF Foundation.

Lehmann, M., Arici, G., Boger, S., Martinez-Pardo, C., Krueger, F., Schneider, M., Carrière-Pradal, B., & Schou, D. (2019). *Pulse of the fashion industry: 2019 update.* Global Fashion Agenda, Boston Consulting Group, and Sustainable Apparel Coalition. <http://globalfashionagenda.com/Pulse-2019-Update> Accessed 20.03.20.

Lehamnn, M., Arici, G., Boger, S., Martinez-Pardo, C., Krueger, F., Schneider, M., Carrière-Pradal, B., & Schou, D. (2020). *CEO agenda 2020: eight sustainability priorities for the fashion industry.* <https://globalfashionagenda.com/ceo-agenda-2020> Accessed 20.03.20.

Myers, W. (2012). *Biodesign: Nature, science, creativity*. London: Thames and Hudson.

Pawlyn, M. (2019). In L. Crook, We fooled ourselves that sustainability was getting us where we needed to go says Michael Pawlyn of Architects Declare. *Dezeen*.

Pennisi, E. (2010). Synthetic genome brings new life to bacterium. *Science, 328*, 958–959. Available from https://science.sciencemag.org/content/sci/328/5981/958.full.pdf, Accessed 27.03.20.

Raber, L. (2010). Nylon's 75th anniversary fete. *Chemical & Engineering News, 88*(15), 46.

Rhodes, M. (2015, February 12). The North Face's 'Moon Parka' is spun from faux spider silk. *Wired*. <https://www.wired.com/2015/12/the-north-faces-moon-parka-is-spun-from-faux-spider-silk> Accessed 04–03–2020.

Rognoli, V., Bianchini, M., Maffei, S., & Karana, E. (2015). DIY materials. *Materials & Design, 86*, 692–702.

Shaikh, A., Chaudhari, S., & Varma, A. (2012). Viscose rayon: A legendary development in the manmade textile. *International Journal of Engineering Research and Applications, 2*(5), 675–680.

Shepherd, H. (2019). Thirsty for fashion? How organic cotton delivers in a water stressed world. *Soil Association*. <https://www.soilassociation.org/thirsty-for-fashion> Accessed 10.09.20.

St. Clair, C. (2018). *The golden thread: How fabric changed history*. London: John Murray Publishers.

St. Pierre, L. (2015). Nature's systems. In K. Fletcher, & M. Tham (Eds.), *Routledge handbook of sustainability and fashion* (pp. 33–42). Routledge.

Sterlacci, F., & Arbuckle, J. (2017). *Historical dictionary of the fashion industry* (2nd ed.). Lanham: Rowman & Littlefield.

Textile Exchange (2018a). *Preferred fiber and materials market report 2018*. <https://textileexchange.org/downloads> Accessed 21.03.20.

Textile Exchange (2018c). *Quick guide to biosynthetics*. <https://textileexchange.org/downloads> Accessed 21.03.20.

The Index Project: Project Coelicolor (2019). <https://theindexproject.org/award/winnersandfinalists/project-coelicolor-body> Accessed 04.03.20.

Vincent, J. (2014). Biomimetic materials. In E. Karana, O. Pedgley, & V. Rognoli (Eds.), *Materials experience* (pp. 235–246). Oxford: Butterworth-Heinemann.

Webster, K. (2017). *The circular economy: A wealth of flows* (2nd ed.). Ellen MacArthur Foundation.

Wijkman, A., & Rockström, J. (2012). *Bankrupting nature. Denying our planetary boundaries*. Routledge.

Worldometer. (2020). https://www.worldometers.info/world-population/world-population-by-year Accessed 10.11.19.

CHAPTER 9

Defining the DIY-Materials approach

Valentina Rognoli[a], Camilo Ayala-Garcia[b]
[a]Design Department, Politecnico di Milano, Italy,
[b]Department of Design, Universidad de los Andes, Bogotá, Colombia

The practices that shape the do-it-yourself (DIY) approach have always considered different sectors of knowledge and experience (Lukens, 2013). The DIY movement is expanding beyond artifacts to include materials from which products are made; namely, DIY-Materials. Designers from all over the world are engaged in various experimental journeys in the field of materials development, and they consider these experiments as the starting point of their design process, which will lead them to the creation of new artifacts. The possibility to self-produce their own materials provides designers with a unique tool to combine unusual languages and innovative design solutions with authentic and meaningful materials experiences.

As this phenomenon of self-production of materials has spread widely in recent years and is starting to be considered as an essential phase of the design process (Bak-Andersen, 2018; Karana et al., 2015; Parisi et al., 2016), it is necessary to investigate and understand it accurately. This chapter aims to provide an updated and comprehensive definition of the DIY-Materials phenomenon, as one of the emerging experiences in the field of design.

Nowadays, and for at least centuries beforehand, the artifacts of daily human life originate mainly from industrial materials, that is, materials developed to meet the requirements and constraints of mass production. Human beings are aware that this model of creation and consumption has led to a problematic situation from an environmental perspective, putting the climate in crisis, depleting nonrenewable resources, creating hazardous waste, and perpetuating an inefficient use of energy resources (Ulluwishewa, 2014).

However, over the last few years, another interesting phenomenon has emerged, bringing the relationship between design, science, technology, production processes, sources, and materials to a new dimension. The phenomenon refers to the rise and use of DIY-Materials, a "new class" of materials, which are conceived by the designer and are characterized by a tinkering

Materials Experience 2. DOI: https://doi.org/10.1016/B978-0-12-819244-3.00010-7

approach and self-production processes (Ayala-Garcia, 2019; Ayala-Garcia et al., 2017; Parisi & Rognoli, 2017; Parisi et al., 2017; Rognoli et al., 2015; Rognoli & Parisi, 2021). This new approach to materials development is enhanced by the renaissance of craftsmanship, by the democratization of technologies, and by combining making, crafting, and personal fabrication practices (Bettiol & Micelli, 2014; Tanenbaum et al., 2013).

9.1 The DIY-Materials phenomenon

Scholars already framed and defined the emerging phenomenon of self-produced materials with a first definition of DIY-Materials in 2015 (Rognoli et al., 2015). The definition describes the materials as self-produced by designers, who follow their own design inspiration while looking for original and unusual sources and adopting a low-cost approach and processes. In short, based on tinkering, this practice guarantees to obtain some material drafts or material demonstrators (Rognoli & Parisi, 2021) with which it is possible to discover and explore the design space of materials.

The phenomenon of DIY-Materials is ubiquitous today, and there are many examples of materials self-produced by designers after intense exploration and tinkering activity.[1] Around 10 years ago, designers began in earnest to create and self-produce materials as if finding immediately ready ones in material libraries was no longer sufficient. For over 20 years, material libraries had been fulfilling their function of bringing closer together designers and the world of materials which, in this case, were industrial materials. Industrial materials are the result of mass production, that is of workmanship of certainty (Pye, 1968), to be chosen considering their expressive-sensorial aspect or to be selected considering their engineering capabilities. The materials included in material libraries' collections are usually materials already developed, marketed, and branded: in other words, ready to be processed and incorporated into a product.

Some designers who embark on a path of development and self-production of materials want to demonstrate their dissatisfaction with the monotonous uniformity of the industrial material landscape, and therefore they try to generate original material experiences, even transforming themselves into real activists[2] (Ribul, 2013) against the mass-production system. They want to take a risk, the workmanship of risk, in which the quality of the result is not predetermined (Pye, 1968). It seems that designers enjoy a renewed synergy between ideation and production processes, getting their hands dirty by

[1] Looking at Dezeen's blog where there are many examples of DIY-Materials developed by professional designers and above all, the DIY-Materials research group's web page: https://www.diy-materials.com/.
[2] http://www.materialactivism.com.

experimenting with colors, textures, consistencies, mixing various ingredients, and having fun looking for alternative and unconventional sources as raw materials.

With DIY-Materials, designers decide not to look at materials that someone has already designed and developed, but instead they inquire/challenge themselves by experimenting and thinking about their own design of materials. This phenomenon does not represent the first time designers have wanted to challenge themselves with the design of matter. In Italy in the 1980s and 1990s, research relating to "The Material of Invention" (Manzini, 1986) and "Neolite" (Manzini & Petrillo, 1991) opened the material creation path, where designers started to include materials as a crucial and decisive moment within the wider design process. DIY-Materials have also strongly influenced research in the domain of materials for design, which today looks with interest at the experimental field of material development. A significant example of this is the recently concluded MaDe (Material Designers) project (2018–20). MaDe was cofunded by the Creative Europe Programme of The European Union and coordinated by Elisava,[3] Politecnico di Milano,[4] and Matter,[5] with the aim to boost talents toward circular economies across Europe. The project focused on the organization of competitions, workshops, an event series, and a web platform, devoted to demonstrating the positive impact material designers can have across all creative sectors.[6]

Another valuable example in this renewed line of research on design materials is the Chemarts project, developed by the School of Arts of Aalto University, Finland. This aimed at innovatively researching bio-based materials while trying to create new material concepts with the help and support of the School of Chemical Engineering of the same University (Kaariainen et al., 2020). A great wealth of studies and research on material design is ongoing,[7] including not only DIY-Materials but also the development of living or growing materials (Ginsberg et al., 2017; Karana et al., 2018; Oxman,

---

[3] http://www.elisava.net/en/research/know-us.

[4] http://www.dipartimentodesign.polimi.it/en/.

[5] https://ma-tt-er.org/.

[6] "Material Designers are agents of change. They can design, redesign, reform, reuse, and redefine materials giving them an entirely new purpose. Increasing the potential of materials, they can go on to research, advise, educate, and communicate what materials are and can be in the immediate, near, and far future, implementing positive social, economical, political, and environmental change across all sectors toward a responsibly designed future." http://www.materialdesigners.org.

[7] Among others are the following: The research group on DIY-Materials based at the Design Department, Politecnico di Milano and the University of Los Andes, Bogotá, Colombia; Professor Carlo Santulli of the University of Camerino, Italy; Professor Carla Langella of the Vanvitelli University of Naples, in collaboration with Professor Mario Malinconico of the CNR of Naples, Italy (http://www.ipcb.cnr.it/index.php/it/personale/strutturato/117-mario-malinconico); and The Royal College of Art's Materials Science Research Center, led by Professor Sharon Baurley.

2020; Zhou et al., 2021), where the designer codesigns together with organisms. Experiments on materials in close collaboration with biologists and scientists are counted among DIY-Materials and expand their definition as they are based on the same process of experimentation and tinkering and on the creation of drafts and demonstrators. Many case studies[8] (Franklin & Till, 2018; Karana, 2020) can be found and substantial research is ongoing in different international universities.[9] Moreover, tinkering and material experimentation are also used frequently in design for interaction, using the hybridization of DIY-Materials with technology and designing and self-producing interactive, connected, and smart materials.[10]

Finally the emergence of the DIY approach to materials is closely intertwined with materials education in the field of design. In many universities and design schools[11] materials are again at the center of the organization's didactic activities, as it was in the first design school, the Bauhaus (Wick, 2000). As Albers stated (1928): "We do not always create 'works of art,' but rather experiments; it is not our ambition to fill museums: we are gathering 'experience'." The emerging new design courses combine the fundamental ingredients of experimentation, a transdisciplinary approach and speculative design, aiming to give the design student tools to approach the world of matter not as a mere observer, but as a participant who can precisely design and self-produce their own materials.

9.2 DIY-Materials: theoretical foundations

As previously mentioned, the concept of DIY-Materials was born following the observation of what was happening in the context of materials for design. This was done using the lens of materials experience, which defines the "experience that people have with and through the materials of a product" (Karana et al., 2014, 2015). DIY-Materials, in fact, lead to a completely different material experience compared to other materials, with regard to the

---

[8] Mogu, https://mogu.bio/; Solaga, http://www.solaga.de/en/; Other cases can be found at http://www.futurematerialsbank.com/.

[9] Among others, there are Professor Elvin Karana, Materials Incubator, TUDelft and Avans Hogeschool, http://www.materialincubator.com/about; Professor Carole Collet, Design and Living Systems Lab, UAL, http://www.designandlivingsystems.com; Professor Carla Langella, Hybrid Design Lab, Università Vanvitelli Napoli, http://www.hybriddesignlab.org/designers-in-lab.

[10] Among others, there are ICS Materials research, Design Department Polimi, http://www.icsmaterials.polimi.it/; Institute for Material Design (IMD) and Professor Markus Holzbach, https://www.hfg-offenbach.de/en/pages/institute-for-materialdesign-imd#about; and Materiability, Professor Manuel Kretzer, Materiability Lab at the University Campus Dessau, http://materiability.com/.

[11] Charlotte Asbjørn Sörensen at School of Art and Communication, Malmö University; Ziyu Zhou at Design Department, Politecnico di Milano; Professor Camilo Ayala-Garcia at Universidad de Los Andes; and Professor Jimena Alarcon at Bío-Bío University.

user who will interact with original materials mainly for their expressive-sensorial component, as well as for the designer themselves. Concerning the designer's role, the main change concerns her/his involvement, as she/he now operates at the forefront of material creation, defining qualities that determine the sensoriality, meaning, and emotional dimension, and the performance of the material (Giaccardi & Karana, 2015).

Compared to what happened in the past, today, designers find a favorable context for their experimental approach to materials because other phenomena have emerged and developed. The first promoting factor is undoubtedly the spread of Fab-labs and maker culture. Following this drive, designers decided to reject the "sit back and be told" school culture (Gauntlett, 2011) and embraced instead making and creating, the idea of continuous works-in-progress, continual experiments with a practical approach to materials, and a curious mind oriented to autonomous research. In addition to maker culture, approaches to frugal innovation were also reevaluated. In specific parts of the world these are called Jugaad (Radjou et al., 2012) or Technological Disobedience[12] (Oroza, 2012; Oroza & de Bozzi, 2002; Rognoli & Oroza, 2015).

The explicit combination of frugality and innovation is relatively new (Albert, 2019). In this way, it is possible to highlight the flexible mentality of those who want to question the status quo, opposing structured approaches, and responding quickly to requests to change the context. Not only it is useful to "think outside the box," but it becomes inevitable to establish new ones.

Designers interested in the material dimension can thus be called Lay Materials Designers following the logic with which Angus Donald Campbell coined the term *lay designer* (Campbell, 2017). The enhancement of these approaches to innovation is made possible thanks to changes that are taking place in the general esthetic language. In fact, today, more than ever, consumers are aware and sometimes require something that is not the result of industrial production, which appears increasingly cold and flat. Everything Standard looks outdated (Friedman, 2010), while imperfection offers an escape (Karana et al., 2017; Ostuzzi et al., 2011a, 2011b; Pedgley et al., 2018; Rognoli & Karana, 2014).

More than in standard industrial production, the designer begins to look with renewed interest into craftsmanship, reevaluating and rediscovering a more intense relationship between the individual and her/his practice. The craftsperson's making places emphasis on visual, tactile, and emotional qualities and represents the ability to convey meaning through form and

[12] http://www.technologicaldisobedience.com.

sensitivity to materials (Yair et al., 1999). Compared to designing on paper and then producing industrially or choosing a material from a library or modeling it on the computer, reevaluating craftsmanship means reevaluating the bodily experience and the relationship that the designer establishes with the material during the making process. As argued by Aktaş & Mäkelä, (2019), as a material becomes a medium to reflect the embodied knowledge, it actively informs the making and conveys information. So the material represents an active nonhuman participant that is no longer passive during ideation. In this way an artifact comes into being through a negotiation process in which both the material designer and the material itself have essential roles. The designer who embraces craftsmanship can become an active maker of materials (Barati & Karana, 2019), capable of exploiting her/his own body to comprehensively explore the potential of materials. The most important tool she/he has at her/his disposal is the hands. As argued by Sennett (2008), Focillon (1939), and Pallasmaa (2009), a design that excludes the hands also disables a specific type of relational intelligence because "making is thinking." When the mind and the hands are divorced, it is the mind that suffers. The hand is the outlet of human will, choice, and action; it characterizes the human being, and it is through the hand that the designer carries out her/his own creative process.

Undoubtedly, craft as a way of thinking through materials (Nimkulrat, 2012) is already included in the context of design research as a way to regain possession of the material culture that surrounds people's daily life. Everything, from furniture to clothes, from food to material tools, can become a field of experimentation because the continuous commitment to make, transform, and modify things and materials can lead to greater awareness and increased autonomy and individual choice (Micelli, 2011). Doing things and making materials—the essential activities of DIY-Materials—make designers more aware of potentials and more able to think radically and unconventionally. The underlying mindset facilitates and embraces interpretation of the incomplete as a positive event of intellectual activity. It is a concrete yet malleable stimulus that cannot be substituted by the simulation and facilitated manipulation of already complete and resolved objects.

For this reason, research on Open-ended Design (OeD) is being developed (Ostuzzi, 2017; Ostuzzi et al., 2017; Ostuzzi & Rognoli, 2019). OeD is the approach by which a result of a product design can change in response to a changed or evolved context. It is defined as suboptimal, error-friendly, unfinished, and contextual, embracing context-dependent errors and is characterized by its internal flexibility, incomplete voluntary definition of its characteristics, and significant imperfection. Making things and self-producing materials, therefore, means reappropriating the delegation that more or less consciously a century ago was granted to large manufacturing companies, separating having ideas from making, and distancing the craftsperson/designer from the pleasure and satisfaction inherent in making (Gauntlett, 2011).

The principle that DIY-Materials practices promote a form of knowing in action (Schön, 1984), that is, experiential knowledge with and through materials, must be incorporated into a broader examination. The outcome of this process is often a self-produced material as a result of making by hand, but this also reflects a dialog and underlying thinking that is induced through the manipulated materials (Nimkulrat, 2012). Thus the process of creating materials by hand can be identified as a way of thinking intellectually (Sennett, 2008) and a way of tinkering manually (Parisi et al., 2017). Furthermore, it satisfies designers' needs for a dynamic process of learning and understanding through material and sensory experience (Gray & Burnett, 2009).

Other significant aspects that distinguish and define today's DIY-Materials and material designers compared to what has been done in the past are undoubtedly the possibility of sharing experiences, the ease of information retrieval, and the democratization of technological practices. Importantly, material research is not limited to self-exploration, as multidisciplinary knowledge sharing platforms are also contributing to the dissemination of materials knowledge to the public (Fadzli et al., 2017). The DIY-Materials approach is configured as a creative activity that implies creation, sharing, and collaboration. As Gauntlett (2011) stated, making is connecting. Nowadays, product design can be practiced at home or as a hobby, supported by the democratization of technological practices (Tanenbaum et al., 2013) that make available commonly used production labs and low-cost equipment and facilities for fabrication.

The last important aspect that must be emphasized about the practice of DIY-Materials is the relationship with sustainability. Today, sustainability is a matter of great importance for society and for the future of Earth and all its inhabitants. For this reason, sustainability is an essential issue within design activity and for designers who have responsibility for shaping our futures. It is stated that more than 80% of the environmental impact of any artifact is determined in the design phase[13] (Thackara, 2006) because during design, decisions are made in relation to production, material usage, and acceptable longevity. A major motivation for designers undertaking DIY-Materials is to find more sustainable and eco-friendly material solutions. However, using a DIY-Materials approach does not guarantee a sustainable material outcome at the end of experimentation. Instead, it is necessary to

[13] "Eighty percent of the environmental impact of the products, services, and infrastructures around us is determined at the design stage. Design decisions shape the processes behind the products we use, the materials and energy required to make them, the ways we operate them on a daily basis, and what happens to them when we no longer need them" (Thackara, 2006, p. 1).

consider that this type of approach leads the designer to make reasonings and choices to acquire a sensitivity toward sustainability challenges since she/he becomes increasingly aware of her/his role as a facilitator and pursuer of sustainable solutions.

The environmental impact of DIY-Materials has not yet been studied in-depth, much less demonstrated. What can be said here is that positive connections can be glimpsed between DIY-Materials and environmental sustainability. Thinking of an alternative approach to the development of design materials can contribute to a more sustainable world and contribute to the three pillars of environmental, social, and economic sustainability (Purvis et al., 2019).

Scholars describe a positive relationship between DIY-Materials and sustainability and identify the potential of the DIY-Materials approach for finding more sustainable material solutions (Alarcón & Llorens, 2018; Fadzli et al., 2017; Karana et al., 2017; Rognoli et al., 2017, 2021; Santulli et al., 2019). Following the reasoning that Martin Albert (2019) makes regarding the relationship between sustainability and frugal innovation, it is possible to elaborate some similar considerations concerning the relationship between sustainability and DIY-Materials.

9.2.1 Environmental sustainability pillar

Developing materials following the DIY-Materials approach is consistent with improving potential ideal models to create green materials, that is, materials that are eco-friendly and useful for concretizing ecological ideas. A DIY-Material is usually developed from the perspective of employing fewer resources, reusing resources, and/or being economically frugal. Habitually, resources are conserved, saved, reduced, and consumed less, while overall more sustainable, renewable, and local resources and processes are preferred. In general the designer who undertakes self-produced materials seeks to improve/maximize energy and material efficiency. Furthermore, sufficiency is an essential part of the environmental sustainability of DIY-Materials. Moreover, the self-produced materials approach creates value from waste (waste as a resource), where existing components and materials are reused, and recycling is performed instead of new material sourcing. The material designer tries to minimize environmental impact by reducing complexity and creating simple, implementable, and repeatable materials and/or material recipes.

9.2.2 Social sustainability pillar

The potential of self-production of materials also consists of realizing social sustainability (Rognoli et al., 2017). The utilization of local resources and local processes could alleviate and reduce global poverty by creating and offering jobs, enabling new entrepreneurship linked to sustainable materials production. Through the development of DIY-Materials, agricultural waste markets are created and valorized as a resource and raw material. As mentioned above, another essential issue concerns better access to information, global knowledge, shared knowledge networks, education and training through DIY-Materials and self-production practices. The DIY-Materials approach promotes the development of social awareness because such materials need to be narrated and described within local contexts and among various stakeholders.

9.2.3 Economic sustainability pillar

Concerning economic sustainability, DIY-Materials appear in general as an opportunity to investigate alternative routes for economic growth that may be used for competitive advantage or to survive material crises in a commercial context. As far as developing countries are concerned, adopting more eco-friendly material development approaches can deliver economic growth according to more sustainable models. The material designer is aware that generating ideas for more sustainable materials can present customers with unique value propositions, potentially increasing competitiveness, improving sustainability performance, and leading to greater productivity and reduced cost per unit related to procurement, production, and distribution. The reason for the positive correlation between economic sustainability and DIY-Materials is intrinsic innovation. Innovation can be exploited commercially and produce, when possible, positive economic effects.

All these introductory observations are necessary for a complete comprehension of DIY-Materials and must be taken into account during the material development.

9.3 DIY-Materials cases: collection and classification

In the last 5 years, DIY-Materials creation has accelerated. More than 100 examples have been collected and examined by the researchers. The collection includes materials created through individual or collaborative self-production experiences, often by techniques and processes of the designer's own invention. Most of these materials are the result of a process of tinkering with materials. The DIY-Materials considered for inclusion in the collection met the criterion of being complete in every part. In particular, great importance was given to the demonstration and video, or photographic documentation, of the experimentation process to create the material.

A fundamental aspect in the definition of self-produced materials concerns precisely the need that the material designer feels to tell the story and she/he does it employing different techniques (storyboard, photos, videos, storytelling, sketches) (Figs. 9.1 and 9.2). The scientist's traditional laboratory notebook is flanked by all kinds of graphic drawings that enrich the idea's communication, allow the sharing and replication of the experimentation and support the acceptability by future users who, through history, can understand precisely the background of the development of the material. As Lambert and Speed (2017) state, even if sharing and distributing processes are commonplace today, the use of narrative brings an additional dimension

(A)

(B)

FIGURE 9.1
Typical layout of DIY-Materials development. *Photo by Jonas Edvard.*

(A)

(B)

FIGURE 9.2
Videos and photography to elaborate the DIY-Materials narrative. *Photo by Jonas Edvard and Sanne Visser.*

to the outcome—the final material or materialized artifact—while increasing the currency in the process itself. It brings people who encounter the material closer to its making, by providing further insight into, and reflection on, our material world. In the DIY-Materials narrative the making is inextricable from the resulting artifact.

Initially the collected DIY-Materials case studies were divided between "DIY new materials" and "DIY new identities for conventional materials." For "DIY new materials," the materials had to be developed starting from the creative use of various substances as material ingredients (e.g., a material made of dried, blended waste citrus peel combined with natural binders). The "DIY new identities for conventional materials" were focused on new production techniques, which rose to new identities for existing materials (they do not necessarily contain new ingredients, e.g., 3D-printed metal and recycled thermoplastics). However, this splitting into two elementary groups was no longer sufficient, given the large number of DIY-Materials examples subsequently found and analyzed from blogs, magazines, books, or created through dedicated workshops. A need to classify DIY-Materials more accurately and more comprehensively has emerged.

As Ashby & Johnson (2002) explained, classification is the first step in bringing order to any scientific endeavor. A classification divides an initially disordered population into groups that have significant similarities, presenting an order. This happened originally in scientific disciplines such as biology, zoology, and geology, contributing to the creation of traditional and still-used classification systems. Materials are classified based on ISO (International Organization of Standardization) descriptors. They are divided into classes (metals, polymers, ceramics, etc.) reflecting similar atomic structures that affect various properties (especially mechanical) and behavior. A correct definition of DIY-Materials must also include the positioning of the materials in a class recognizable by scientific communities. However, given the experimental nature of DIY-Materials and their sometimes mixed ingredients, the ISO classifications of materials are rarely sufficient or relevant.

Furthermore, scholars have already highlighted how, for artistic disciplines, standard classifications of materials are inadequate or unsatisfactory for communicating material qualities and information useful for design (Ashby, 2011; Marschallek & Jacobsen, 2020). In particular, Marschallek & Jacobsen (2020) elaborated an alternative and systematic way of defining and classifying what is meant by the term "material substances."

Given these premises, the activity of collecting DIY-Materials case studies focused on finding what could be described as the distinctive feature of the materials. This was identified as the sources from which the whole creative process starts. Sources are defined as "a place, person, or thing from which something originates or can be obtained."[14] In this context the term "source" indicates the origin and is different from the word "resource," which means "source of supply or support";[15] in short, something that serves as an aid, especially one that can be readily drawn upon when needed. From careful observation and analysis of the material stories created by designers to report and document their tinkering and experimentation activities, it was deduced that the sources were a fundamental element capturing the material designer's great attention. The material storytelling always starts from the description of the source and the reasons that prompted the designer to consider it.

DIY-Materials are developed using different material substances as ingredients, for which the "main" ingredient constitutes the source that determines the origin of that particular material. All the collected DIY-Materials case studies were then grouped according to the source used to start the creative process, and families of individual materials with similar origins were created. The taxonomic categories used for DIY-Materials are called "Kingdoms" and are distinguished based on the source used by the material designer to start the design process focused on materials (Ayala-Garcia, 2019; Ayala-Garcia et al., 2017).

9.4 DIY-Materials classification: the five kingdoms

The categorization phase involved analyzing the 100 DIY-Materials case studies based on their material source. As has already been said, the source is an essential element in the DIY-Materials approach. Based on the sources of the 100 case studies, five principal families (kingdoms) were deemed necessary. As the first three families shared an exact analogy with the biological classification of natural elements conceived by the Swedish botanist, zoologist, and physician Carolus Linnaeus called *Systema Naturae*,[16] it was decided to proceed to develop the biological metaphor by completing the Systema with missing families.

The Systema Naturae was for several years the standard biological classification of the elements of the Earth. Known as Linnaeus' taxonomy,[17] this

---

[14] https://dictionary.cambridge.org/dictionary/english/source.

[15] https://www.dictionary.com/browse/resource.

[16] https://en.wikipedia.org/wiki/Systema_Naturae.

[17] Carl Linnaeus (1707–78) laid the foundations for modern biological nomenclature, now regulated by the Nomenclature Codes, in 1735. He distinguished two kingdoms of living things: Regnum Animale ("animal kingdom") and Regnum Vegetabile ("vegetable kingdom," for plants). Linnaeus also included minerals in his classification system, placing them in a third kingdom (not living), Regnum Lapideum.

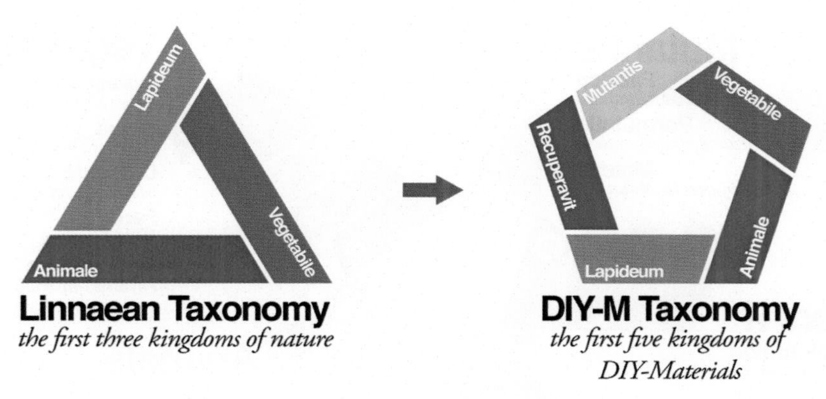

FIGURE 9.3
New kingdoms emerged for DIY-Materials classification as a result of contemporary human impact.

historical publication established a hierarchical classification of the world's natural elements, dividing it into three main kingdoms: animal, plant, and mineral.

Although modern biological terminology has evolved into many deeper divisions than the taxonomy of nature, Linnaeus's original approach provided an excellent and engaging basis for naming the DIY-Materials kingdoms.

To the three original kingdoms: animals, plants, minerals (kingdoms Vegetabile, Animale, and Lapideum), two others have been added. The first concerns sources deriving from waste and scraps (kingdom Recuperavit) and the second relates to information technologies and interaction (kingdom Mutantis) (Fig. 9.3). Hence the kingdoms mainly obey the source of the material substance, which becomes the main ingredient for developing the self-produced material. Classes, as subcategories, were then generated for each kingdom based on further similarities identified between the sources within each kingdom (Fig. 9.4).

9.4.1 Kingdom Vegetabile

The first kingdom is the "kingdom Vegetabile." It refers to every material development where the primary source derives from plants and fungi.[18]

DIY-Materials developed under this kingdom differ from the others, mainly because they can be originated from growing or farming techniques.

[18] In Linnaean taxonomy, the classes and orders of plants, according to his Systema Sexuale, were never intended to represent natural groups. Classis XXIV. Cryptogamia included the "flowerless" plants such as ferns, fungi, algae, and bryophytes.

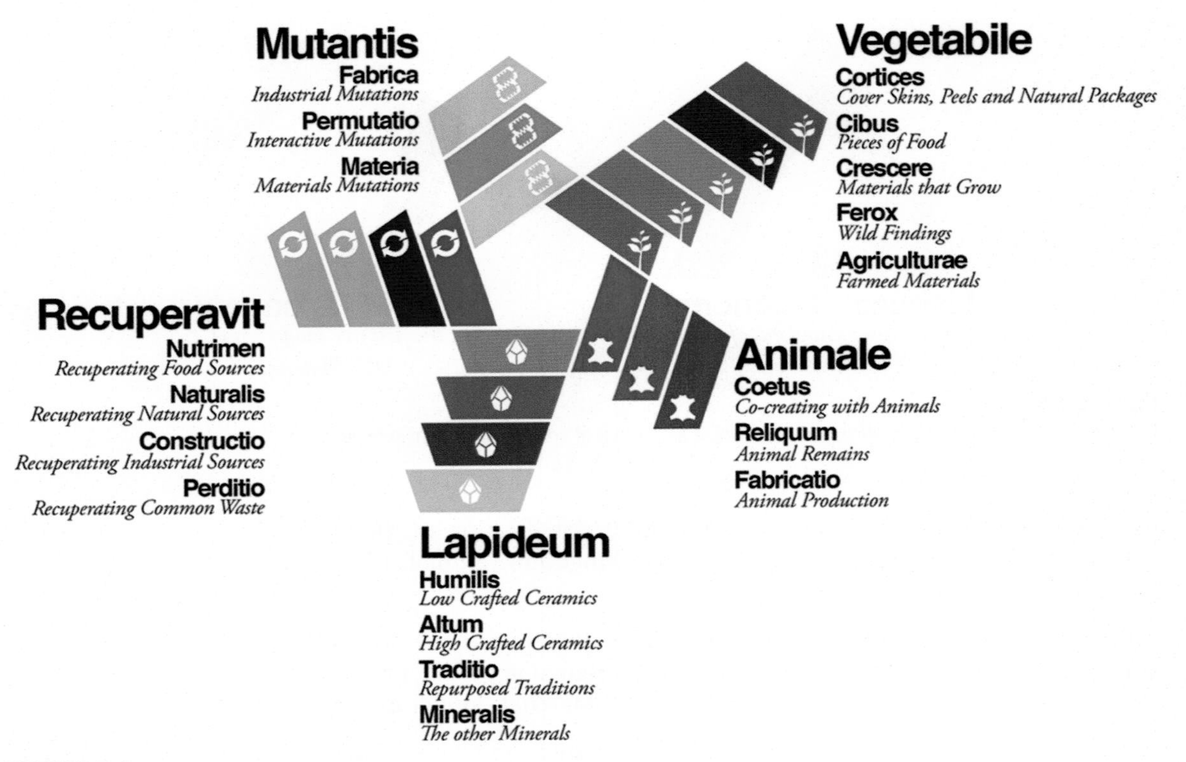

FIGURE 9.4
The DIY-Materials kingdoms with their classes.

Designers who create materials under this category collaborate with, for example, farmers and biologists. Kingdom Vegetabile is subdivided into (Fig. 9.5) *Farmed Materials; Wild Findings; Materials that Grow; Pieces of Food; Cover Skins, Peels, and Natural Packages*.

9.4.1.1 Farmed materials (Agriculturae)

This class includes all the different materials obtained using a growing crop or a domesticated plant as a source. It covers all the DIY-Materials created as alternatives to industrial production with traditional means of agriculture, respecting the timing and seasons and maintaining a biological cycle that allows each application to be compostable or degradable in the medium/short term.

9.4.1.2 Wild findings (Ferox)

Since some DIY-Materials are obtained from agricultural systems, others can be created by picking wild sources. Therefore the source is

Vegetabile

Agriculturae
Farmed Materials

GreeNet* By Aversa, Ertin, Wang, Bettoni.
A material developed using celery fibers. By twisting filament, the celery can be knitted.

Leek Paper*
by Erdogan Fumagalli Paillieux, Hasegawa.
obtained by peeling the surface layers of the leek and inked when dry.

Ferox
Wild Findings

Story of a Pinetree by Sarmite Polakova.
From the inner bark of the Pine, leather-like material with natural color, elegant and robust character.

Catching Autumn*
by Bektaş Kettl, Aydin & Milly.
Different types of leaves in different colors with wax finishing.

Crescere
Materials that Grow

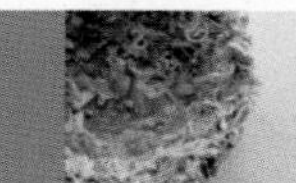
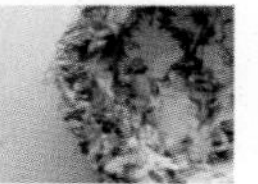

A Matter of Time* by Stefano Parisi.
Mycelium, the fast-growing, vegetative part of fungi with flax seeds.

EC Grass*
by Milanese, Kagaine, Park & Jimenez.
Made of pure grass. Light structures combining length of fibers.

Cibus
Pieces of Food

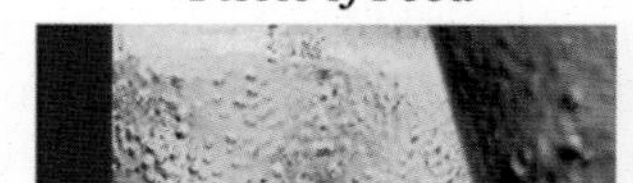

Sugar Glass by Fernando Laposse.
Glazing, an old tradition in the patisserie industry replicable to produce rotational molding shapes.

Rice* by Xu Mengdi.
Material made out of rice, one of the biggest food leftovers in the world. Mixed different kinds of rice is possible to obtain various colors and textures.

Cortices
Skins, Peels, Natural Package

Piñatex by Carmen Hijosa.
leaf pinneaple fibers to produce a non-woven mesh with a similar leather-like appearance.

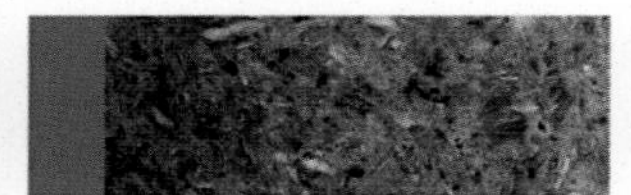

Artichair by Spyros Kizis.
Artichoke thistle mixed with a bio-resin to produce sort of plant fiber reinforced polymer.

FIGURE 9.5
Examples of DIY-Materials belonging to kingdom Vegetabile.

characterized by the geographical area from which it is collected, such as the rainforest, tropical environment, grasslands, or coniferous forests. Designers collect sources that are not considered necessary by industrial counterparts because they are not reliable as a supply, difficult to obtain, or hard to scale. However, they remain excellent sources for the realization of local DIY-Materials, favoring zero-kilometer economies and circular approaches.

9.4.1.3 Materials that grow (Crescere)

Growth is a natural phenomenon characterized by a living organism's ability, or part of it, to develop and increase its volume over time. Materials in this subcategory are recognizable as the designer often, with the aid of scientists, grows some of these organisms in controlled environments. This class includes fungi, algae, and plants. The vast majority of these organisms pose no real threat to humans, plants, or animals; therefore they are studied as an opportunity or alternative source to be used as substitutes to produce more sustainable materials.

9.4.1.4 Pieces of food (Cibus)

The development of materials from sources that help the sustenance of communities is a very sensitive issue. However, some substances suitable for feeding humans can be contraindicated if consumed in large quantities, in which case they can be used for other purposes, such as DIY-Materials development. In this subcategory, finding materials in which the primary source is an edible or recognizable nutrient is possible.

9.4.1.5 Cover skins, peels, and natural packages (Cortices)

The materials that belong to this subcategory are developed starting from natural skins, peels, or nutshells in which fruits and vegetables are biologically protected. When these items are removed and thrown away, they could be used as a source to self-produce materials employed for short-life applications before being discarded.

9.4.2 Kingdom Animale

The "Animal Kingdom" refers to all material sources derived from animals. Bacteria have also been included as a possible source in this kingdom, although they are not considered to belong to the animal kingdom in modern biological classifications (Schleifer, 2009). DIY-Materials of this kingdom can be developed using parts (hair, skin, bones, etc.) of animals or through the collaboration of animals or microorganisms such as bacteria. Kingdom Animale is classified into (Fig. 9.6) *Cocreating with Animals, Animal Remains,* and *Animal Production.*

9.4.2.1 Cocreating with animals (Coetus)

In addition to getting daily sustenance, some animals work to create artifacts useful to guaranteeing their own life as much as sourcing or protecting food. There are many examples in the world of insects (e.g., bees, ants, worms, termites, and spiders), but other species can also build and process materials to create, for example, a nest, a dam, or a coral reef. In the history of humankind, nature has always been an essential reference for conceiving innovative

Animale

Coetus
Co-creating with Animals

From Insects by Marlene Huissod.
Made from extracted propolis bio resin from the beehive, carefully heated and sculpted.

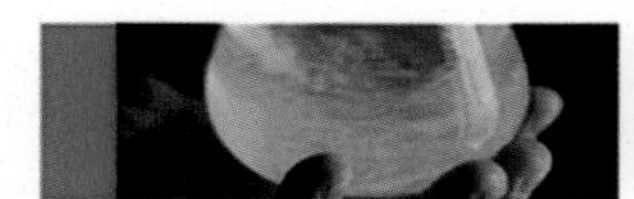

Performative Bioluminiscence by Bahareh Barati.
exploring bio-luminescent bacteria as an interactive element in daily objects.

Reliquum
Animal Remains

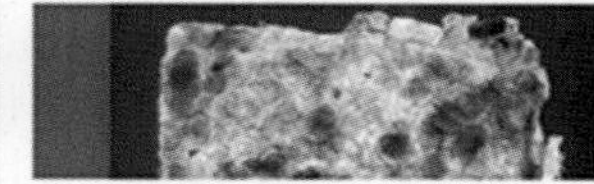

Fish Left (L)over* by Claudia Catalani.
Made out of unedible parts of fishes. Different mixes and elements produce transluscent and light materials.

Porcaria*
by Machado da Silva, Ren, Simon, Manlin.
Material developed to change pigskin's properties and appearance.

Fabricatio
Animal Production

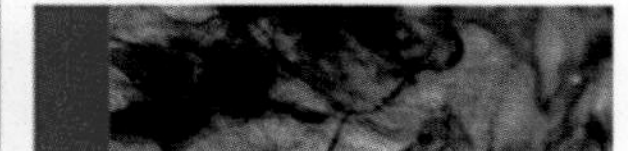

Gelawool*
by Bell, Muñoz, Winandi & Queirolo.
a flexible non-stretch material made out of gelatin with Alpaca wool.

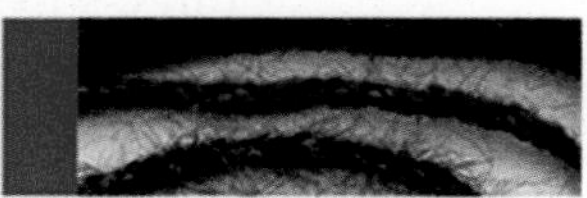

Tricology by Sanne Visser.
Human hair as raw material to create ropes and utilitarian objects.

FIGURE 9.6
Examples of DIY-Materials belonging to kingdom Animale.

artifacts. Numerous advances in technology, design, and art have arisen in the broad basin of biological inspiration. In recent years the bioinspired approach has become very popular (Benyus, 1997), and in addition to specific projects, a large body of research is developing.[19] Regarding DIY-Materials, animals' behaviors and abilities can be considered sources for cocreating materials and even artifacts.

Some designers analyze animal behaviors and, without interfering with the biological means that such actions represent for the sustenance of a particular species, have initiated a process of cocreation. In this class of DIY-Materials, it is possible to find materials in which animals address the primary source through the guidance of an artificial structure or a material manipulated by a designer after being processed by another organism. In this class, much more is experienced in the process, sharing it with the animal, rather than material sources.

9.4.2.2 Animal remains (Reliquum)

Life and death are, in general, a delicate subject to be addressed and themes quite far from the context of design materials. Problems related to overpopulation of the planet and access to food derived from intensive farming are

[19] https://biomimicry.org/.

putting pressure on the environment and compromising human beings' future. The call to action is necessary. This class of DIY-Materials belonging to the animal kingdom considers as a source the enormous quantity of carcasses and shells of animals that are discarded after extracting their nutrients for human sustenance. This mass begins to weigh negatively on public health. Many different solutions are under discussion, starting from consuming less meat to the diversification of the human diet. Nevertheless, a challenge that can be faced by designers is how to use the remaining parts of the animals usefully and creatively. The materials in this class are recognizable as the designer creatively uses leftovers from organic animals to compose structures, sheets, or volumes.

9.4.2.3 Animal production (Fabricatio)

Compared to other animal species, human beings can project and therefore can see design opportunities in the natural phenomena they are surrounded by. For example, humans have always bred animals for their own sustenance, taking advantage of everything that could be used and grown back (e.g., sheep's wool. Many organic elements capable of regeneration can be used as material sources. The materials in this class include everything provided and created by animals, including humans, which grow, can be cut, cleaned, and creatively processed.

9.4.3 Kingdom Lapideum

The kingdom of the minerals was the third kingdom recognized by Linnaeus and related to the first two (plants and animals), it was the only kingdom encompassing nonliving elements. The kingdom Lapideum includes all the DIY-Materials from minerals such as stones, sand, pottery, and clay. In some cases, sources belonging to other kingdoms, such as wool or cotton fabrics, were combined with sources from kingdom Lapideum (but in a lower percentage than the primary material substance).

Kingdom Lapideum is divided into (Fig. 9.7) *Low Crafted Ceramics, High Crafted Ceramics, Repurposed Traditions*, and *Other Minerals*.

9.4.3.1 Low crafted ceramics (Humilis)

By definition, craftsmanship deals with details and does not allow itself to be distracted from them to consider mass production needs. The craftsperson focuses on transforming materials and creating unique pieces, enhancing, for example, characteristic and unique imperfections or defects, such as a knot in wood or a strip in stone. The DIY-Materials developed and belonging to this class are recognizable by the designer's intention to enhance traditional craftsmanship, mainly with ceramics as a source. Other stones and terracotta may be included, as long as the designer maintains conventional ways of working with the material. In this class, much more is experienced in the process rather than on material sources.

Lapideum

Humilis

Low crafted ceramics

Balloon Bowls by Maarten De Ceulaer.
Clay with a polymeric hardener inside two air-inflated balloons.

Improvisation Machine by Annika Frye.
A DIY rotational molding process in which polyper plaster is poured.

Altum

High Crafted Ceramics

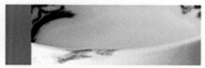

Marwoolus® by Marco Guazzini.
A combination of marble dust with pieces of wool and resin obtaining a colored marble material.

Ceramic wood by IMD Offenbach.
A ceramic made out of wood pieces. The obtained shape retains the natural form of the plant.

Traditio

Repurposed Traditions

Blueware Vases by Studio Glithero.
Ceramic finish by exposing elements to ultraviolet light using the early principles of photography.

Kivi by Erin Turkoglu.
Hammering clay of different colors into a mold.

Mineralis

The Other Minerals

Melach*
by Grubisič, Nikaein, Zhou & Shabani.
Material made from different types of salt.

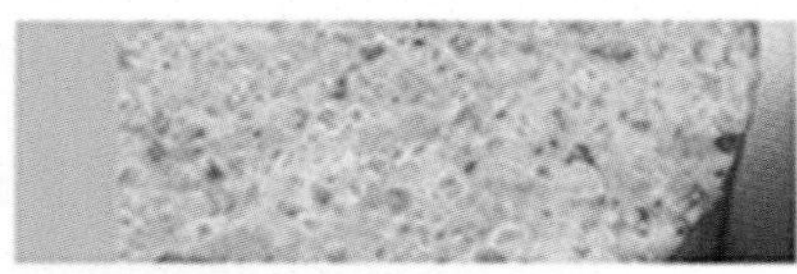

Nabatea* by Talithashani, Cui, Chanalithichai & Chen.
Pink Himalayan salt with a colored concrete appearance.

FIGURE 9.7
Examples of DIY-Materials belonging to kingdom Lapideum.

9.4.3.2 High crafted ceramics (Altum)

New craftsmanship and modern artisans with an industrial backend can achieve exciting results, thanks to the democratization of technology. While the means of modern production is moving toward automation and data exchange in manufacturing, through what is known as Industry 4.0, the traditional machines for industrial production that have become the mainstream of engineering development over the past century are more accessible than ever. A new army of tech artisans is starting to produce materials by gaining access to industrial infrastructure or hacking machines and controlling their behavior with mere coding knowledge. In this class, it is possible to find ceramic and composite materials that can be configured with a particular automated technology aid. The peculiarity of the DIY-Materials in this class lies in the process rather than in the recipes.

9.4.3.3 Repurposed traditions (Traditio)

Contemporary society's technological advances improve the way something is done by changing the manufacturing technique or introducing a different approach to achieve a similar result. It happens to many other objects and products made in a particular way such as shoes, bags, glass vases, or lamps. For many years, skilled artisans built those elements, but due to the inevitable need to increase production, either the tradition was abandoned or replaced by a new efficient technique.

The materials in this class are recognizable as the designer reproposes old or traditional methods to create a particular element. Photography, engraving, illustration, or other techniques are often used to provide a particular finish to the skin of the ceramic material. It is an essential strategy as it provides the designer with rich tools and techniques to achieve unexpected results.

9.4.3.4 The other minerals (Mineralis)

In this class are grouped some impressive developments of stone materials that are not related to traditional minerals. Curiosity and fearless experimentation, decisive components of the designer's mentality, reveal other material development sources that are sometimes not so evident. This class of material sources spans unexpected substances such as dust, salt, or crystals that create unique compositions of materials.

9.4.4 Kingdom Recuperavit

The kingdom Recuperavit is the first that was introduced specifically for the classification of DIY-Materials. This kingdom includes all materials self-produced from sources that the external context considers as a waste, but which have potential to be transformed into precious resources. In the words of Michael Ashby (2013) "Waste is waste only if nothing can be done to make it useful." DIY-Materials under this kingdom often come from plastic, metal, or organic waste, sometimes as a by-product of industrial production or sometimes due to the throw-away society. At present, this is the most powerful kingdom, considering the number of cases observed. Within this realm the designer's intention toward a more conscious and sustainable future is explicit. Kingdom Recuperavit has the following classification (Fig. 9.8): *Recuperating Food Sources, Recuperating Natural Sources, Recuperating Industrial Sources*, and *Recuperating Common Waste*.

9.4.4.1 Recuperating food sources (Nutrimen)

Tackling food "waste" is primarily an ethical question. Therefore definite critical issues emerge about the overproduction of food, its industrial production and transformation, as well as the impact of its packaging and transport. In this class, it is possible to find different materials originated using organic waste and leftovers as sources that designers creatively transform by converting them into resources. The waste/sources considered here concern both the domestic sector, those of shared consumption of commercial establishments, and large-scale distribution up to and including the level of industrial waste.

9.4.4.2 Recuperating natural sources (Naturalis)

Natural sources are not necessarily a problem. Unlike food overproduction, various material self-development opportunities can be found by recovering materials from a wide range of sources. Once again the designer's trained eye is what plays a crucial role in identifying resources that can transform into something significant before returning to the ground and could replace inorganic materials in a wide range of critical applications. In the book *Cradle to Cradle* (McDonough & Braungart, 2002) the authors criticize the common desire for perfection and geometric management of natural resources such as grass or garden bushes. The authors criticize modern society's unnatural behavior: everything should look fantastic, pointing out how much time and energy it takes to remove every inch of grass and bushes, fighting against natural plant growth. The materials in this class are recognizable as the designer is making innovative use of different biological sources which are often considered useless by society. For instance, elements like corn stalk and leaves sometimes are discarded during the process of collecting corn to feed humans and animals. Such sources, when creatively used by designers, can become an important source to replace common industrial materials and foster social innovation (Rodriguez, 2017).

9.4.4.3 Recuperating industrial sources (Constructio)

As discussed above, industry produces enormous leftovers, which are unsuitable for feeding animals and organisms, or not put to use in any meaningful way for consumption. They should be kept away from any

Recuperavit

Nutrimen

Recuperating Food Sources

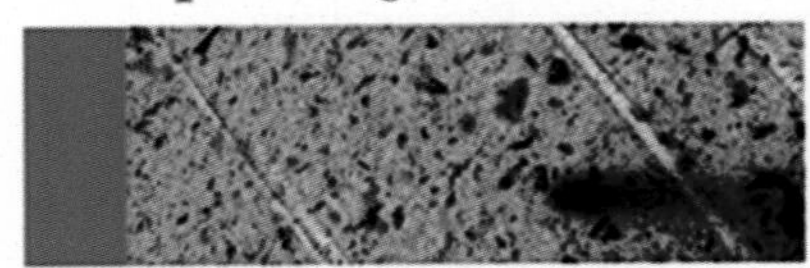

Fruit Leather Rotterdam.
Semi-transluscent leather like material made out of rotten fruits from local markets.

Decafé by Raúl Laurí.
Coffee grounds mixed with a natural binder, pressed and heated to obtain a solid material.

Naturalis

Recuperating Natural Sources

Funco* by Catalani, Chen, Panza & Yang.
Cotton pads used to remove makeup from the skin. It is colored in different ways depending on the makeup color.

Cornstark DIY-M by Karen Rodriguez.
Discarded Cornstalk and leaves from the agricultural production of corn.

Constructio

Recuperating Industrial Sources

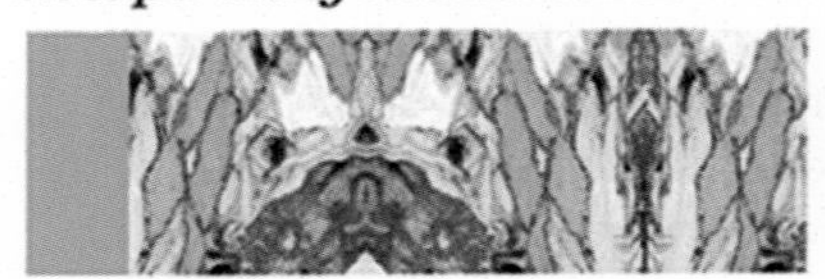

Structural Skin by Jorge Penadés.
Scraps from leather production with bone glue. A solid structural material with good isolation properties.

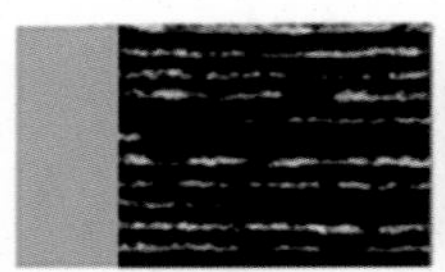

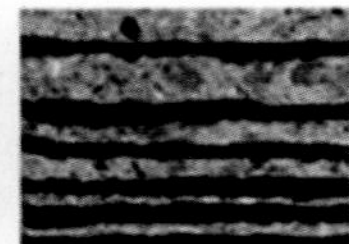

Re-Surface* by Helga Aversa.
Recuperated fiberglass from abandoned boats and yatchs. The material can be develpoed with different patterns and colors.

Perditio

Recuperating Common Waste

Fluff* by Koski, Munda, Kruger & Salehi.
An elastic and semi-transparent material made out of the remaining particles inside the drying machine's filter.

Butts Bunny* by Studio Swine.
A semi-translucent material with a certain degree of flexibility made out of discarded cigarrette butts.

FIGURE 9.8
Examples of DIY-Materials belonging to kingdom Recuperavit.

natural ecosystems to avoid damage. Many industries reuse their scraps and have an industrial closed metabolism system (McDonough & Braungart, 2002), but many others do not. In several countries where regulations and legislation are not as strict as they should be, even large companies avoid investing in waste management as it is not mandatory. Fortunately, designers are more aware of becoming material hunters in such urban settings. Materials in this class are recognizable as the designer identifies a useful source from an industrial leftover, unused components, or a product entering a final stage of its life cycle.

9.4.4.4 Recuperating common waste (Perditio)

Throughout the kingdom Recuperavit, different sources are classified according to their natural or industrial origin. However, there are some elements of unselected provenance, which may belong to another class.

Common waste is a term used to describe all unsorted daily discarded waste. It comes in such quantities in different places that it can become almost invisible (Freinkel, 2011). For example, buying food packaged in plastic is so common that people do not calculate the package's shelf life. Plastic does not hurt, and it is not worth waging open warfare like has been happening in recent years. The plastic is not at fault; it is humans' failure to deal with plastic sensibly and sensitively that is the matter. As designers, consumers, or industrialists, plastic is relied upon as a commodity, the implications of which on a large scale, regarding disposal, are severe. Many such sources are problematic as they can include different material substances in a single composition, making them difficult to separate at the end of their life cycle. In this class it is possible to find several alternatives to treat this unselected material from untraceable sources of origin.

9.4.5 Kingdom Mutantis

The kingdom Mutantis is the second newly added kingdom, to fully define the DIY-Materials phenomenon. This realm includes DIY-Materials created from different technological mixes and the hybridization of industrial, interactive, or intelligent sources (Ritter, 2007). In this category it is possible to see combinations of different material sources that come from another realm but evolve into something particular with the aid of technology. This transformation represents a significant change in the nature and behavior of the material compared to other kingdoms. Kingdom Mutantis is divided into (Fig. 9.9) *Industrial Mutations*, *Interactive Mutations*, and *Material Mutations*.

9.4.5.1 Industrial mutations (Fabrica)

"Hacking" is a word with its origins in the German term "Hacken." It is an adequate definition to explain how to intervene, disassemble, or form a

Mutantis

Fabrica
Industrial Mutations

Mx3D by Joris Laarman.
An industrial multiple axis robot with a welding machine to produce 3D metal shapes.

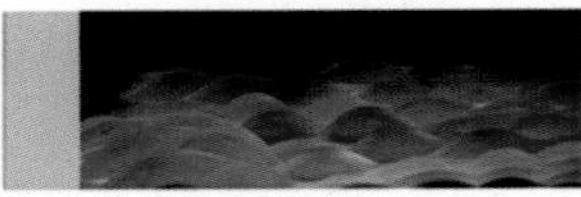

Maku by Valentin Brück.
3D printed pneumatic system.
Microorganisms are printed in a special nutrient solution in liquid silicone.

Permutatio
Interactive Mutations

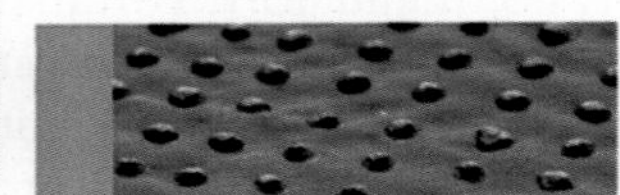

Magnetic Fabrics by Lilian Dedio.
Arrangement of magnetic components inside a textile allowing movement of the material with the aid of electronics.

Enactive Materials by K. Franinovic.
Electroactive polymers with responsive behavior to explore agency, materiality and interaction.

Materia
Material Mutations

Paralight Skin by Pohlman, Schwarze and Wöhrlin.
Silicone combined with leather that respond to touch and pressure.

GLAS_hybrids by IMD Offenbach.
Direct handling of different materials to find new relationships and shapes.

FIGURE 9.9
Examples of DIY-Materials belonging to Kingdom Mutantis.

particular structure or system into pieces. In the domain of materials, hackers are the same artisans and bricoleurs, who learn by taking apart and transforming, changing the principal behavior of a material, machine, or a particular tool, to obtain a result. Designers who have access to different tools and machinery can create improvements and change the devices' behavior by introducing new techniques. This process leads to the creation of new materials and subsequently to new products with original languages. The materials in this class are recognizable as the designer creates a material by intervening on a particular type of machinery or transforming and hacking a specific process. For this specific class the source is the process.

9.4.5.2 Interactive mutations (Permutatio)

The field of interaction design has provided essential tools for intervening on the behavior of materials through programming and controlling properties and qualities. Interaction has become relevant in design as it has highlighted the importance of focusing on behaviors of things and people (Cooper et al., 2007; Moggridge, 2007). The advent of open-source printed circuit boards, which began with Arduino, Wiring, or Raspberry several years ago, allowed designers to control a material's behavior and activate interactions between people and the material. It is still an approach to material design that is in an embryonic stage. Some technological advances may be necessary before electronic circuits and computing capabilities can be incorporated into a material, and they are considered the real sources in this case.[20] However, several designers are manipulating material properties with the help of these tools and envisioning alternative futures. Materials in this class are recognizable as the designer fuses any standard material with a computational layer that exhibits a desirable property that can be controlled and produce a specific performance.

9.4.5.3 Material mutations (Materia Mutantur)

Material hybrids and composites are one of the four categories of the so-called engineering materials. Inside this category it is possible to find a whole universe of materials composed of two principal elements with specific properties or characteristics to add to the composite: the matrix and the reinforcements. Sometimes they can include a third element called the core. A material composite aims to obtain an improved system with augmented properties compared with the original matrix material. Recently, designers have embarked on similar quests for hybridization and compositions of two or more elements. The aim is usually a mix of seeking superior performance and sensorial improvements. Materials in this subcategory are recognizable as the designer mixes two or more base materials, different from composites, creating new material languages embodying future applications that challenge the current state-of-the-art.

9.5 Interrelationships between kingdoms and their updated definition

The purpose of this chapter was to provide an updated and comprehensive definition of the DIY-Materials phenomenon, as one of the emerging experiences in the field of design. Categorization of DIY-Materials has been approached from the perspective of identifying the various sources that form the origin for DIY-Materials development. The categorization is made according to "kingdoms," through which the numerous development potentials of DIY-Materials can be conveniently highlighted. Furthermore, the kingdoms help indicate the designers of materials the many material substances that can be used as a source. However, it should be emphasized that, despite the classification introduced here, the sources and consequent DIY-Materials can belong to several kingdoms at the same time. In other words the boundaries between categorizations can be quite flexible when developing DIY-

[20] In these circumstances, we should mention the concepts of Computational Composites (Valgarda & Redstrom, 2007), Smart Materials Composites (Barati et al., 2019), ICS Materials (Parisi et al., 2018), and Open Materials (by Catarina Mota, http://openmaterials.org/).

Materials. Furthermore, the classification is of great help to identify the different types of DIY-Materials produced and to guide their design and development.

Through the classification, it is possible to understand how any given material in hand can belong to two kingdoms, build a bridge between categories, and enrich the definition of a material that shows advantages, characteristics, and properties from each side of the categorization. Projects such as Fruit Leather[21] or Marwoolus[22] demonstrate this principle well. The first of these materials belongs to the kingdom Recuperavit based on its foundations to recover food sources that are thrown away daily in various urban markets. However, given its condition as a natural plant source, it could also be considered a breeding material from the kingdom Vegetabile. The second material is part of the family of high craftsmanship ceramics within the kingdom Lapideum since the material emerges by setting the industrial machinery to achieve the designer's will. Simultaneously, it is a material produced by intelligently recovering the leftovers of two industries from the same region by mixing them into a new material, also inscribing the kingdom Recuperavit.

In light of what has been said and explained, it is now possible to generate a more complete and updated definition of DIY-Materials to conclude the discussions presented in this chapter.

The original definition of DIY-Materials (Rognoli et al., 2015) highlighted a new practice whereby designers create materials using DIY methods, providing an opportunity to reconsider the characteristics of existing production processes alongside the properties of industrialized materials, for the purpose of developing new unique experiential qualities. In the original definition the practice of material-making was understood to be an individual or collective endeavor, requiring the designer to work with suitable tools for experimentation and the production of the material. In recent years the DIY-Materials phenomenon has expanded strongly, requiring more refined ways of describing the material-making processes and classifying the created materials. After having carried out in-depth research on the phenomenon, formulating a theoretical basis to collect, analyze, and classify the different cases, the following extension to the original definition is proposed.

DIY-Materials are defined by sources of matter that may be located in one of five material kingdoms. DIY-Materials differ from other materials when the transformation of an unconventional source is evident (e.g., cultivated vegetables, animal constituents, raw minerals, and recovered waste). The

[21] https://fruitleather.nl/.

[22] http://marcoguazzini.com/marwoolus/.

motivations behind the development of DIY-Materials are powerful and mainly concern the designer's desire to look for alternatives to the world of standardized industrial materials. Furthermore, the designer's will is also to experiment and to undertake research so that their design path concentrates on the material itself and not only on its application. DIY-Materials designers have a desire to go back to handling and designing materials and to have influence right at the start of the process.

To develop these materials, designers embrace available technologies and work within material creation studios, spaces, and labs. The DIY-Materials approach is transdisciplinary and interdisciplinary. In other words, it is an approach that at the same time goes beyond and intertwines different disciplines, essentially rejecting the fragmentation of knowledge, aiming instead at an integrated and unitary understanding of the phenomenon. To this end, the DIY-Materials development practice often involves biologists, materials scientists, environmental engineers, and other specialists outside the core of "design." DIY-Materials are born from experimentation with an unconventional source. In the earliest phases of development the designer typically experiments with low-tech tools and equipment, often conceiving and producing these tools directly. Such experimentation is essential to learning material possibilities and is called tinkering.

The initial phase of DIY-Materials creation is therefore called "Tinkering with Materials" (Rognoli & Parisi, 2021), supporting the acquisition of knowledge on the sources and developing procedural understanding. Here the designer aims to obtain information and understand the qualities of sources and their empirical properties, recognizing their constraints and identifying their potential. Tinkering promotes sensory awareness of material attributes and can reveal unpredictable and unique results. Tinkering with materials means working with the hands and the direct involvement of all human senses. Through this practice, the possibilities of how materials can look, feel, sound, and smell are discovered, and it is possible to manipulate the sources to create "material drafts." The second phase of DIY-Materials development is called "Tinkering for Materials," focusing on the creation of "material demonstrators." The material demonstrators are designed and delivered as the outcome of an experimentation process. The most common material demonstrators are those aiming to explore and represent quality variants for DIY-Materials such as color, thickness, and texture. There are also demonstrators of processes, that is, showing variations around the creation of forms. Demonstrators emerging from "Tinkering for Materials" become a valuable resource for design activity. In fact, by tinkering for materials without a design application in mind, the designer adopts exploratory research to create and nurture a vision that may further develop the material or its meaningful application.

Both the proofs and the demonstrators are, however, potential materials, which cannot be considered as finalized but rather a work-in-progress. Their general qualities of being affordable and competitive, and crucially still open to modification and development, make them attractive as tools for speculation and reasoning about possibility and potential. DIY-Materials can help solve a specific problem, by creating and embodying material visions to understand and evaluate which path to take.

An essential aspect in the definition of DIY-Materials is the creative narrative (Lambert & Speed, 2017). The material ideation and development must be communicated. This is frequently achieved by accompanying the material creation process with the main tools of storytelling, such as videos, storyboards, and diaries. This narrative component is an integral part of the DIY-Materials phenomenon and is fundamental for communicating the perception and vision of the new proposed material, as well as defining the material identity and its future acceptance in the lives of people who may come in contact with or use the material.

In closing a quick final consideration needs to be made on the issue of scaling-up DIY-Materials creation. Material demonstrators have the potential to be taken as a model for companies to replicate and refine the materials from the perspective of commercial application, especially locally as an alternative to mass production and supply of industrial materials. At the academic level, it is exciting to show how this type of approach is relevant for the training and professional practice of the designer who becomes a material designer, possessing the sensitivity and propensity to design not only forms but also material qualities. It is also crucial to highlight how this approach can become an opportunity for companies interested in more sustainable material alternatives and attentive to their local economies because, in fact, it would allow the improvement of alternative and more flexible production systems even at an industrial scale.

Acknowledgments

All cases presented in this chapter marked with an asterisk (*) in teh images, were developed under the made@polimi workshops or Master's thesis at Politecnico di Milano and are available at https://www.diy-materials.com/. The authors are grateful to Owain Pedgley for his feedback in the preparation of this chapter.

References

Aktaş, B. M., & Mäkelä, M. (2019). Negotiation between the maker and material: Observation on material interactions in Felting Studio. *International Journal of Design, 13*(2), 55–67.

Alarcón, J., & Llorens, A. (2018). DIY Materials and circular economy: A case study, educating industrial designers for sustainability. In *Preprints 2018*.

Albers, J. (1968). Creative education. In H. M. Wingler (Ed.), *The Bauhaus: Weimar, Dessau, Berlin, Chicago* (p. 142). Cambridge, MA: The MIT Press.

Albert, M. (2019). Sustainable frugal innovation. The connection between frugal innovation and sustainability. *Journal of Cleaner Production, 237*, 117747.

Ashby, M. F. (2011). *Materials selection in mechanical design*. Oxford: Butterworth Heinemann.

Ashby, M. F. (2013). *Materials and the environment* (2nd ed.). Oxford: Butterworth Heinemann.

Ashby, M. F., & Johnson, K. (2002). *Materials and design, the art and science of materials selection in product design*. Oxford: Butterworth Heinemann.

Ayala-Garcia, C. (2019). *The Materials Generation. The emerging materials experience of DIY-Materials* [Unpublished Ph.D. thesis in Design, Design Department, Politecnico di Milano] funded by Colombian Government and Universidad de Los Andes.

Ayala-Garcia, C., Rognoli, V., & Karana, E. (2017). Five kingdoms of DIY-Materials for Design. In *Proceedings of international conference 2017 of the Design Research Society Special Interest Group on Experiential Knowledge (EKSIG)* (pp. 222–234), Delft University of Technology Het Nieuwe Instituut, Rotterdam, The Netherlands, 19–20 June 2017.

Bak-Andersen, M. (2018). When matter leads to form: Material driven design for sustainability. *Temes Disseny, 34*, 12–33.

Barati, B., & Karana, E. (2019). Affordances as materials potential: What design can do for materials development. *International Journal of Design, 13*(3), 105–123.

Barati, B., Karana, E., & Hekkert, P. (2019). Prototyping materials experience: Towards a shared understanding of underdeveloped smart material composites. *International Journal of Design, 13*(2), 21–38.

Benyus, J. M. (1997). *Biomimicry: Innovation inspired by nature*. William Morrow & Co.

Bettiol, M., & Micelli, S. (2014). The hidden side of design: The relevance of artisanship. *Design Issues, 30*(1), 7–18.

Campbell, A. D. (2017). Lay designers: Grassroots innovation for appropriate change. *Design Issues, 33*(1), 30–47.

Cooper, A., Reimann, R., Cronin, D. (2007). *About Face 3: The Essentials of Interaction Design*. Indianapolis, Indiana: Wiley.

Fadzli, I. B., Aurisicchio, M.,& Baxter, W. (2017). Sustainable materials in design projects. In *Proceedings of International Conference 2017 of the Design Research Society Special Interest Group on Experiential Knowledge (EKSIG)* (pp. 194–207). Delft University of Technology Het Nieuwe Instituut, Rotterdam, The Netherlands, 19–20 June 2017.

Focillon, H. (1939). *In praise of hands*. Parkstone Press.

Franklin, K. A., & Till, C. A. (2018). *Radical matter: Rethinking materials for a sustainable future*. London: Thames & Hudson.

Freinkel, S. (2011). *Plastic: A toxic love story*. Boston: Houghton Mifflin Harcourt.

Friedman, T. L. (2010, October 25). Average is over. *International Herald Tribune*.

Gauntlett, D. (2011). *Making is connecting. The social meaning of creativity, from DIY and knitting to YouTube and Web 2.0*. Polity Press.

Giaccardi, E., & Karana, E. (2015). Foundations of materials experience: An approach for HCI. In *Proceedings of the SIGCHI conference on human factors in computing systems* – CHI'15. New York: ACM Press.

Ginsberg, A., Calvert, J., Schyfter, P., Elfick, A., & Endy, D. (2017). *Synthetic aesthetics: Investigating synthetic biology's designs on nature*. MIT Press.

Gray, C., & Burnett, G. (2009). Making Sense: an exploration of ways of knowing generated through practice and reflection in craft. In *Proceedings of 'Crafticulation and Education' International Conference of Craft Science and Craft Education*, University of Helsinki, Finland. Helsinki University Press. *Journal of Design, 12*(2), pp. 119–136.

Kaariainen, P., Tervinen, L., Tapani, V., & Riutta, N. (2020). *The Chemarts cookbook*. Aalto Arts Book.

Karana, E. (2020). *Still alive: Livingness as a material quality in design*. Open Source Publication. Available from https://issuu.com/caradt/docs/still_alive_caradt_avans_vweb.

Karana, E., Barati, B., Rognoli, V., & Zeeuw Van Der Laan, A. (2015). Material driven design (MDD): A method to design for material experiences. *International Journal of Design, 19*(2), 35–54.

Karana, E., Blauwhoff, D., Hultink, E. J., & Camere, S. (2018). When the material grows: A case study on designing (with) mycelium-based materials. *International Journal of Design, 12*(2), 119–136.

Karana, E., Giaccardi, E., & Rognoli, V. (2017). Materially yours. In J. Chapman (Ed.), *The Routledge handbook of sustainable product design*. London: Routledge.

Karana, E., Pedgley, O., & Rognoli, V. (2014). *Materials experience: Fundamentals of materials and design*. Elsevier.

Karana, E., Pedgley, O., & Rognoli, V. (2015). On materials experience. *Design Issues, 31*(3), 16–27.

Lambert, I., & Speed, C. (2017). Making as growth: Narratives in materials and process. *Design Issues, 33*(3).

Lukens, J. (2013). DIY infrastructure and the scope of design practice. *Design Issues, 29*(3), 14–27.

Manzini, E. (1986). *The material of invention*. MIT Press.

Manzini, E., & Petrillo, A. (1991). *Neolite-Metamorfosi delle plastiche*. Milano: Edizioni Domus Academy.

Marschallek, B. E., & Jacobsen, T. (2020). Classification of material substances: Introducing a standards-based approach. *Materials and Design, 193*.

McDonough, W., & Braungart, M. (2002). *Cradle to cradle: Remaking the way we make things*. North Point Press.

Micelli, S. (2011). *Futuro artigiano: L'innovazione nelle mani degli italiani* [Future craftsman: Innovation in the hands of the Italians]. Venezia: Marsilio Editori.

Moggridge, B. (2007). *Designing Interactions*. Cambridge, MA: The MIT Press.

Nimkulrat, N. (2012). Hands-on intellect: Integrating craft practice into design research. *International Journal of Design, 6*(3), 1–14.

Oroza, E. (2012). Technological disobedience. *Makeshift—A Journal of Hidden Creativity, 3*(1), 50–53. Available from http://mkshft.org/technological-disobedience/.

Oroza, E., & de Bozzi, P. (2002). *Objets Réinventé. La création populaire à Cuba*. Paris: Editions Alternatives.

Ostuzzi, F. (2017). *Open-ended design. Explorative studies on how to intentionally support change by designing with imperfection* [Unpublished Ph.D. thesis in Design, Ghent University, Faculty of Engineering and Architecture].

Ostuzzi, F., De Couvreur, L., Detand, J., & Saldien, J. (2017). From Design for One to Open-ended Design. Experiments on understanding how to open-up contextual design solutions. *The Design Journal, 20*(1), S3873–S3883.

Ostuzzi, F., & Rognoli, V. (2019). Open-ended design: Local re-appropriations through imperfection. In *Proceedings of the 3rd LeNS World Distributed—Designing sustainability for all* (Vol. 3), Milano, 5–7 April 2019.

Ostuzzi, F., Salvia, G., Rognoli, V., & Levi, M. (2011a). *Il valore dell'imperfezione: l'approccio wabi-sabi al design [The value of imperfection: The wabi-sabi approach to design]* (pp. 1–135). Milano: Franco Angeli.

Ostuzzi, F., Salvia, G., Rognoli, V., & Levi, M. (2011b). The value of imperfection in industrial products. In *Proceedings of DPPI'11: designing pleasurable products and interfaces*, June 2011. Italy: Politecnico di Milano.

Oxman, N. (2020). *Material ecology*. New York: Museum of Modern Art.

Pallasmaa, J. (2009). *The thinking hand: Existential and embodied wisdom in architecture*. John Wiley & Sons Ltd.

Parisi, S., & Rognoli, V. (2017). Tinkering with Mycelium. A case study. In *Proceedings of the international conference on experiential knowledge and emerging materials, EKSIG 2017—"Alive. Active. Adaptive"* (pp. 66–78). Rotterdam, The Netherlands, 19–20 June.

Parisi, S., Rognoli, V., & Ayala-Garcia, C. (2016). Designing materials experiences through passing of time: Material driven design method applied to mycelium-based composites. In *Proceedings of the 10th international conference on design and emotion, celebration & contemplation* (pp. 239–255), Amsterdam, 27–30 September.

Parisi, S., Rognoli, V., & Sonneveld, M. (2017). Material tinkering. An inspirational approach for experiential learning and envisioning in product design education. *The Design Journal, 20*, S1167–S1184.

Parisi, S., Rognoli, V., Spallazzo, D., & Petrelli, D. (2018). ICS Materials. Towards a re-interpretation of material qualities through interactive, connected, and smart materials. In *Proceedings of DRS international conference 2018, Design as a catalyst for change*, Limerick, Ireland, 25–28 June.

Pedgley, O., Şener, B., Lilley, D., & Bridgens, B. (2018). Embracing material surface imperfections in product design. *International Journal of Design, 12*(3), 21–33.

Purvis, B., Mao, Y., & Robinson, D. (2019). Three pillars of sustainability: In search of conceptual origins. *Sustainability Science, 14*, 681–695.

Pye, D. (1968). *The nature and art of workmanship*. Cambridge University Press.

Radjou, N., Prabhu, J., & Ahuja, S. (2012). *Jugaad innovation: Think frugal, be flexible, generate breakthrough growth*. Jossey-Bass Inc Pub.

Ribul, M. (2013). *Recipes for material activism*. Open-Source Publication. <https://issuu.com/miriamribul/docs/miriam_ribul_recipes_for_material_a>.

Ritter, A. (2007). *Smart materials in architecture, interior architecture and design*. Basel: Birkhäuser.

Rodriguez, K. (2017). *Corn stalk Do-It-Yourself materials for social innovation* [Master's thesis supervised by Valentina Rognoli and Camilo Ayala-Garcia, Politecnico di Milano, Milano]. Retrieved from institutional repository.

Rognoli, V., Ayala-Garcia, C., & Bengo, I. (2017). DIY-Materials as enabling agents of innovative social practices and future social business. In *Proceedings of Diseno conciencia—Encuentro International de disegno, Forma 2017*, La Habana, Cuba.

Rognoli, V., Ayala-Garcia, C., & Pollini, B. (2021). DIY recipes. Ingredients, processes and materials qualities. In L. Cleries, V. Rognoli, S. Solanki, & P. Llorach (Eds.), *Material designers. Boosting talent towards circular economies*. <http://materialdesigners.org/>.

Rognoli, V., Bianchini, M., Maffei, S., & Karana, E. (2015). DIY Materials, Special Issue on Emerging Materials Experience *Materials & Design, 86*, 692–702.

Rognoli, V., & Karana, E. (2014). Towards a new materials aesthetic based on imperfection and graceful ageing. In E. Karana, O. Pedgley, & V. Rognoli (Eds.), *Materials experience: Fundamentals of materials and design* (pp. 145–154). Elsevier: Butterworth Heinemann.

Rognoli, V., & Oroza, E. (June 2015). "Worker, build your own machinery!" A workshop to practice technological disobedience. In *Proceedings of PLATE conference 2015, product lifetimes and the environment*. Nottingham Trent University, UK.

Rognoli, V., & Parisi, S. (2021). Material tinkering and creativity. In: L. Cleries, V. Rognoli, S. Solanki, & P. Llorach (Eds.), *Material designers. Boosting talent towards circular economies.* <http://materialdesigners.org/>.

Santulli, C., Langella, C., & Caliendo, C. (2019). DIY materials from potato skin waste for design. *Journal of Sustainable Design, 3*(3).

Schleifer, K. H. (2009). Classification of bacteria and archaea: Past, present and future. *Systematic and Applied Microbiology, 32*(8), 533–542.

Schon, D. A. (1984). *The reflective practitioner: How professionals think in action.* Basic Books.

Sennett, R. (2008). *The craftsman.* Yale University Press.

Tanenbaum, J.G., Williams, A.M., Desjardins, A., & Tanenbaum, K. (2013). Democratizing technology: Pleasure, utility and expressiveness in DIY and maker practice. In *Proceedings of CHI 2013*, Paris, France, 27 April–2 May 2013.

Thackara, J. (2006). In the Bubble, Designing in a Complex World. MIT Press.

Ulluwishewa, R. (2014). *Environmental unsustainability. Spirituality and sustainable development.* London: Palgrave Macmillan.

Valgarda, A., & Redstrom. (2007). Computational composites. In: *Proceedings of the 2007 conference on human factors in computing systems, CHI 2007*, San Jose, CA, USA, 28 April–3 May 2007.

Wick, R. K. (2000). *Teaching at the Bauhaus.* Hatje Cantz Pub.

Yair, K., Tomes, A., & Press, M. (1999). Design through making: Crafts knowledge as facilitator to collaborative new product development. *Design Studies, 20*, 495–515.

Zhou, J., Barati, B., Wu, J., Scherer, D., & Karana, E. (2021). Digital biofabrication to realize the potentials of plant roots for product design. *Bio-Design and Manufacturing, 4*.

CHAPTER 10

Design and science: a pathway for material design

Carla Langella
Department of Architecture and Industrial Design (DADI), University of Campania "Luigi Vanvitelli", Caserta, Italy

10.1 Intersections between design and science

The relationship between design and materials science can be placed within the wider framework of the scenario built on the intersection of design culture and the sciences. In recent years, more and more designers have chosen to weave their paths with scientific disciplines, offering themselves as an element of connection between scientific knowledge and society. The strengthening of the relationship between design and science is now a widespread phenomenon, a media success, which has acquired multiple forms generated by the combination of infinite design variants and multiple scientific areas (Langella, 2019).

Many of the science and technology achievements have produced revolutions that profoundly affect not only people's way of life but also their opinions and choices. The design of the Anthropocene (Seidl et al., 2013) plays an important role in the processes of decoding and adapting to change since it is capable of giving shape and meaning (Verganti, 2018) to the changes induced by science and technology, along with new knowledge, transmitting them to society through objects, communication devices, services, critical interpretations, and material experiences (Hekkert & Karana, 2014) elaborated with accessible languages, which can allow as many people as possible to know, metabolize and use the results of the techno-scientific evolution in a conscious way (Olson, 2000). For these reasons, in recent decades design—intended both as practice and as thought—has increasingly intersected its paths with the so-called "hard" sciences such as physics, chemistry, mathematics, biology, as well as with the most fluid and interdisciplinary contemporary sciences: synthetic biology (Agapakis, 2013), neuroscience (Zuanon, 2014), bioengineering (Kutz, 2003; Ramsden et al., 2007), and nanotechnology (Kemp, 2017).

Materials Experience 2. DOI: https://doi.org/10.1016/B978-0-12-819244-3.00001-6

The designer who collaborates with science, induced by the intention of enhancing the results for the benefit of society, summarizes the modern narrative vision of science as a vector of development, improvement, growth, and emancipation: placing in it a trust very similar to the cultural modernity approach evoked by Habermas in *Der philosophische Diskurs der Moderne* (Habermas, 1986). However, these current phenomena, like those of the past, occur according to nonlinear development paths, through leaps and round trips, consistent with the flows of contemporary science, which proceeds in the new directions of sustainability, equity, social well-being, and regeneration of the body and mind—building a bridge of continuity with the optimistic-scientific visions of modernity.

An important contribution to the focus on the new forms of relationship between design and science was the exhibition curated by Paola Antonelli *Design and Elastic Mind* held at the MoMA in 2008, the year of the beginning of the large-scale diffusion of this cultural phenomenon. According to Antonelli, designers and scientists should collaborate by making mutual contributions to their paths, establishing biunivocal relationships to prefigure possible futures together (Antonelli, 2008). As Joichi Ito writes, this connection includes both the science of design and the design of science, as well as the dynamic relationship between these two activities (Ito, 2016).

Designers can create connections between new scientific knowledge and people's lives. Their ability to grasp epochal changes in technology, science, and society by transforming them into objects and behaviors that people adopt, more or less consciously, makes it possible to translate the results of contemporary science into new expressive and productive trajectories. At the same time, designers are able to provide science with indications on new directions and research themes to be undertaken that respond to the needs of society and the market as well as any cultural changes or to stimulate the creative capacity of science with points of view and design matrix approaches.

The collaboration between design and science is increasingly frequent both in design schools and in the professional work of designers, especially for younger generations. The diffusion of these phenomena, along with the need to fully define and understand them, resulted in the founding of a journal in 2016: the *Journal of Design and Science* (JoDS), a joint venture between the MIT Media Lab and MIT Press, which aims to investigate the new connections between design and science by breaking down the barriers between traditional academic disciplines, while exploring current and controversial scientific, design, and social issues, with particular attention to mutual interactions. Overcoming the barriers must not be intended as a symptom of homogenization or loss of complexity, but rather as a paradigm within which new forms of connection and intersection between design and science are generated, aimed at developing hybrid products in which different points of view and skills coexist and collaborate (Oxman, 2016). In other words, a paradigm shift profoundly revises project and design culture.

10.2 Divergences and convergences

In the field of material design, there are several reasons that push designers toward science: at times less utilitarian and at times rational reasons. Science can offer inspiration for innovative material solutions, from both a performance and experiential point of view, in a time when stylistic innovation is increasingly difficult and technological innovation is harnessed by homologation. The designer who collaborates with the science of materials can be motivated by the ethical choice to help bring scientific knowledge closer to society, but often also by a desire to legitimize their work through the attribution of the values of utility and indispensability recognized to science.

These reasons can be traced back to the tendency of design to look to science for new material solutions to problems related to the protection of the environment and the well-being of people. In some cases, it seems that designers are looking for an ethical or cultural justification in the science of materials for choosing to propose new products in a market that is now saturated. Assisting science, a vector of innovation and advancement for contemporary society, through the project, to improve the quality of life, can be perceived as a noble intent, capable of rehabilitating the cultural and social image of the design discipline.

In opening *The emergence of a scientific culture: Science and the shaping of modernity* Stephen Gaukroger (2006, p. 11) uses a quote by Nietzsche from *Die Geburt der Tragödie* which states "What is the significance of science viewed as a symptom of life? (...) Is the resolve to be so scientific about everything perhaps a kind of fear of, and escape from, pessimism? A subtle last resort against-truth? and, morally speaking, a sort of cowardice and falseness?" (1872, p. 18). Even if design, interested in making a contribution to the evolution of materials, looks toward the sciences, it does not always act fully aware of the scientific process.

The attempt to use science as a sort of guarantor of meaning does not always correspond, on the part of the designers, to a rigorous knowledge of the scientific issues that characterize the materials with which it approaches and, therefore, the actual implementation potential in product design. For example, it is quite rare (and probably unreasonable to expect) that a designer can design materials at a molecular level like a chemist or to personally carry out assessments related to process parameters. The objective of the intersection with materials science for designers must not be to replace scientists, but rather to propose new points of view, interpretations, strategies, and applications, related to research and scientific discoveries, using the tools of the design culture, linked to the perceptual and experiential aspects, which can even go so far as to contribute to the advancement of scientific culture in the sectors in which they are inserted.

In order for this to happen, the limits and obstacles that resist the cooperation between design and the sciences, which is linked first of all to a distance of the languages and objectives of the research, must be made explicit. One of the main obstacles to a biunivocal relationship is the difficulty of synchronization. Scientific research takes a very long time, based on the experimental method which includes attempts, errors, intuitions, and verifications. For this reason, scientists working in the field of new materials tend to set precise and short-term goals, while trying to optimize their work as much as possible. The main aim is to reach the results before anyone else in order to publish them in the most prestigious magazines as original data, avoiding any deviations and slowdowns.

Design, on the other hand, thanks to new digital modeling and production technologies, is able to shorten, compared to the past, the time required from conception of the project to the verification of the prototypes and production. If the design process does not involve particularly complex processes, it can take a relatively short time, even a few months—unimaginable compared to scientific research, which generally takes years to obtain ascertained and publishable data. For these reasons, designers must be aware of the need to plan short and effective meetings, avoiding digress so as to respect the needs of scientists. In addition, it is useful to quickly identify the needs and effectiveness of the technology transfer of scientists to envisage possible applications from the beginning of the collaboration. The target audience is also very different. Scientists come to new knowledge through original but repeatable protocols, which they communicate to a scientific community in the sector, generally international but very specific and limited. These needs can contrast with the attitude of design to solve design problems by designing characteristic products that can be understood and appreciated by the largest possible market. Unlike the sciences, the scale of intervention of design is much wider and can also address different types of targets since there is no form of selective ranking.

The use of language in materials science is also very precise and specialized, based on strictly binding conventions and rules. Terminological translations, inaccuracies, or personal interpretations are not allowed since they would change the meaning and scientific value of a statement. Terms such as hardness, rheological characteristics, elasticity, and plasticity have very precise meanings linked to the measurable and testable technical properties of the materials. Design, on the other hand, plays with languages in a more flexible and creative way, also using rhetorical tools such as analogies, metaphors, allegories, translations, antitheses, and similarities. Through these modes, it is able to generate original, innovative expressions and syntactic devices, which often deviate from the conventional use of terminologies. When discussing materials, designers refer to perceptual and experiential properties that are difficult to define in an absolute and precise way, but inevitably require nuanced, hybrid, allusive, and evocative languages. This distance must be bridged by designers who must strive to be respectful of the terminological and precise conventions when dealing with scientific topics. At the same time, they must try to transfer their criteria for defining materials in the most shareable way possible, also by expressing them through schemes, infographics, or prototypes that allow them to directly experience the different forms of quality related to experiences.

Furthermore, to reach the production target, the design must necessarily use available and applicable knowledge, materials, and technologies, while adhering to specific economic and regulatory constraints. Science, however, to be considered original, aims to define experimental procedures that must be as new and unexplored as possible. This difference can induce the world of design and production to go beyond the usual boundaries to outline innovative production processes that only with the help of science is it able to catch a glimpse of.

In addition to distances, there are also many affinities that bring design and science closer together. First of all is the common attitude to make research projections converge toward distant but possible and feasible futures. An important prerequisite for any successful collaboration is that common goals are established. To do this, it is important to mediate between the priority objectives of scientists and designers. Scientists are interested in publishing in high-impact journals, as well as obtaining funding. For this reason, it is important to demonstrate the technological transferability of the research results through the Technology Readiness Level (TRL), that is, the maturity and implementability index of new materials or transformation processes developed in a potential production context (Peters, 2014). Design can assist science in prefiguring the best opportunities to translate scientific knowledge on specific materials into production results by increasing the TRL index and, consequently, the opportunities to attract public or private funding. Designers are also interested in realizing and bringing their own projects to the market and people's lives. It is important when starting new collaborations between design and science to highlight this affinity and immediately begin to focus on research and experimentation paths that can facilitate the technological transfer of scientific and design research, with the conversion of material inventions into products and services aimed at end users.

Finally, collaboration with scientists can help design researchers be involved in scientific publications with a higher impact than those of their own sector, while for scientists the possibility of producing patents and technology transfer and third mission operations increases. The awareness of the main differentiating factors and affinity between design and science serves both designers and scientists to make them aware of the individual specificities to induce them to relate, respecting these specificities with a propensity for dialog and the achievement of common objectives. It is important that, from the first moments of the interaction, mutual curiosities, common interests, common testing, and simulation platforms arise that allow to simultaneously verify the technical properties and experiential qualities up to obtaining a shared intent and joint prefigurations of new scenarios and products.

Compared to the past, there are now cultural and instrumental prerequisites to facilitate scientists and designers when collaborating. Multidisciplinarity, interdisciplinarity, and transdisciplinarity are now required by most public funding calls for research, and the concepts of sharing and hybridization of skills are now acquired in the common lexicon of researchers from both the scientific and design fields (Bellotti et al., 2016). For the distances identified to be reduced so as to generate advancements, it is therefore important that the dimension of learning, in the field of both design and science, adapt to these evolutionary processes by offering students the tools and skills necessary to tackle innovation based on scientific research, oriented, and mediated by design.

10.3 The evolution of the relationship between design and materials science

The history of the role of design in the design of materials must be read in the light of the evolutionary scenario of the relationship between design and science. The relationship between designers and traditional materials has been characterized since the Industrial Revolution by an experimental approach that led designers to adapt to the characteristics of the materials that were limited to the available technologies, sometimes trying to push the limits of their properties in an empirical and experimental way, perhaps in collaboration with the manufacturing companies. In the last century, new materials were developed primarily in chemical laboratories, on the basis of many discoveries, such as those made in the macromolecular field that led to the progressive increase in polymeric materials.

In the last few decades of the 20th century, the rapid proliferation of new hyperperforming materials such as technopolymers, superconductors, smart materials, superalloys, and nanomaterials based on the discoveries and scientific inventions of chemists, physicists, and engineers led designers to gradually lose confidence in their ability to manage such materials and control their increasingly complex properties. This marked a departure from the integrated relationship between design and materials that had characterized the design culture of the 1950s.

The creation and diffusion of physical and digital material libraries was supposed to help designers keep up to date and access the information needed to make an informed selection of new materials for their projects. However, the exponential speed with which the number of materials was collected probably increased their feeling of vertigo and inadequacy. The subsequent phenomena related to the *Maker* and *Do It Yourself* (DIY) movements (Anderson, 2012) had the merit of bringing designers closer to experimentation on materials, inducing them to enter into an interspace between research on materials and their application, in a position closer to that of the craft workshop than to that of the scientific laboratory.

Human value is a dimension in materials design that has been recovered. Values related to both the sensory experience and the evocative and imaginative dimensions have acted as an antidote to the anxiety of performance generated by a pressing technological-scientific progress. Among the designers who deal with materials, the category that adopts an empirical approach is the most active and widespread. They are mainly young designers who act in first person on the material compositions through attempts, tests, and experiments—mainly consisting in the original combination of available materials and formulations, which often include waste or natural elements. With this approach, designers sometimes experiment instinctively, others according to a more rigorous plan that approaches a scientific method, resulting in a solution that best approximates their design vision. It is a modality that does not necessarily require the participation of scientists, but which certainly benefits from a basic knowledge of the fundamentals of chemistry as well as the ability to search and study similar experiments in the scientific literature.

Driven by the spread of the Maker phenomenon, these experiments are often very close to the open-source culture which leads to confront, in a mainly informal way, with the dimension of self-production. The materials developed in this context are called *Do-It-Yourself materials* (Rognoli et al., 2015) because they are the result of self-production, and *Open Materials*, when the procedures used to make the materials are shared in open source through communities, generally digital. The maker, open-source and open innovation approaches, based on the principles of cooperation and the innovation potential of communities, makes it possible to disseminate and share experiences with a strong experimental content, making the procedures repeatable and updatable through sharing vehicles such as recipes and tutorials often diffused via the internet.

Designers enter the production process in a new way, albeit mainly at a reduced scale in terms of the quantity of pieces produced, with the production of limited series or single prototype pieces. The interest in the processes leads to the research for a new and deeper relationship with the material and its workability. A relationship becomes a dense experience due to its sensorial, perceptual, and anthropological value. The concept of material experience (Karana et al., 2008) extends beyond the user experience of the final product to also include that of the designers in the phases of conception and manipulation of the new materials. The choice to manipulate the material autonomously and spontaneously becomes an opportunity to investigate the relationship between nature and artifice, along with the issues of sustainability, through the use of materiality rather than dematerialization. Designers are moving toward new renewable, recyclable, recovered, and ennobled materials. For example, they are experimenting with new forms of *upcycling* of waste products or reproposing vegetable resins and fibers which, for technical-economic reasons, had been supplanted by petrochemical materials in the last century. Along with the contemporary recovery of the materials of the past, designers are also reinterpreting the techniques, procedures, and tools found in ancient ethnographic memories, but implementing them with innovative digital tools.

For several years, the self-production approach has been strongly driven both from below through the Maker movement and from above through universities, schools, and design academies, mainly Central European, which have focused on processes, the modification of machinery, the reuse of waste, and the transformation of easily accessible materials, which also includes food ingredients. This has achieved an important paradigm shift carrying a significant value, not so much for the results achieved in terms of newly designed materials but, above all, for the culture of design that regains confidence in its ability to intervene on the material.

The implementation of DIY-materials in design teaching (Parisi et al., 2017) allows young students to become familiar with the physical and concrete component of the project—with the material experience. From empirical experiences, design students learn to experiment with processes and compositions, refining their ability to intuit and prefigure new material scenarios. They approach scientific method acquiring, directly in the field, the awareness of the need to carry out experiments with the rigor of protocols. A rigor leads them to gain experience in the integration of raw materials, processes, material properties, and applications.

It is infrequent that self-produced materials, developed by designers in contexts other than scientific laboratories, translate into the industrial dimension. It is more likely that they can remain in a sphere of limited reproducibility such as that which characterizes some specific applications strongly linked to customization such as orthopedic supports (prostheses, braces, orthoses), which must often be made to measure as in the *Thumbio* project shown in Fig. 10.1 (Caliendo et al., 2018). In an empirical approach to materials, the design is based on aspects related to intuition, synesthetic perception (Schifferstein & Wastiels, 2014), as well as the instinctive and emotional prefiguration of the fruitful experience that constitutes the genetic heritage of the design discipline.

In both research and the profession, the DIY approach is not enough for designers to concretely impact the innovation of materials. Once they have gained greater confidence with formulations, processes, compositions, and an experimental approach, designers must take a further step toward materials science, which is increasingly open to the possibility of cross-fertilization and collaboration with design contexts. What has changed in recent decades is that, while in the past centuries, the development of new materials was based on a discovery approach that strongly limited the participation to chemists, physicists, and engineers; today the research on materials is based on

FIGURE 10.1

Thumbio, bioplastic immobilization brace made with starch-based biomaterial, waste liquid from mozzarella production, and hemp fibers. *Designed for the rhizarthrosis pathology by Clarita Caliendo with the collaboration of material engineer Carlo Santulli and the orthopedic doctor Antonio Bove.*

the integration of knowledge by opening-up to a much wider range of actors that also includes designers. This openness must, however, correspond to an adequate orientation on the part of designers who often do not have the skills or self-confidence necessary to exercise an active role.

10.4 The new material experience generated by the intersection between design and science

To participate in the innovation of materials, designers must necessarily immerse themselves in the techno-scientific dimension of materials, inoculating the reasons for the project, the human values of the sensory and emotional experience, and the visions of a more sustainable future in which materials can adaptively adhere to the new complex needs of contemporary living. The increasing proximity between design and science, along with the openness offered to designers by the integration paradigm to participate in the processes of development and innovation of materials, increases the number of designers who choose to concretize their visions by penetrating into the material dimension with different gradients of awareness and interrelation with scientists.

Since the second decade of the 2000s, driven by a greater awareness of the need to be assisted in their experiments by chemists or materials engineers, along with a greater openness and practice in the interdisciplinary dialog, some designers have intervened on materials in an increasingly more incisive and integrated way with science (Wilkes et al., 2016). These designers no longer simply offer compositions indirectly based on instinct and intuition, but personally experience a direct relationship with the designing of materials from the point of view of science, conducting their activities in scientific laboratories outside of the specific context of their design projects. They discover scientific method, the rigor of experiments, protocols, and rational criteria (Langella, 2019). The commitment to the experimental scientific process becomes the moment in which the designer transfers his visions, but also the most pressing needs of society, to the world of materials.

The critical and operational tools of design allow to enhance the research results by recovering and elevating what can be defined as innovation waste, that is, new materials developed in the laboratory that have yet to find an application. Material scientists, although inclined to develop products and services that enhance their research, often fail to do so due to the lack of time or approach and, for this reason, their experiment risk remaining as such, rather than passing from the scientific research stage to that of innovation and then product. It is in the scientist's own attitude not to stop at the result achieved, but to project himself toward the next one so as not to risk that other researchers get there first. It is this race against time that, in many cases, leaves no room for technology transfer (Kotha et al., 2013). This scenario opens a new gateway to design which, by integrating the results of science with market trends, and with the evolution of contemporary languages, needs, and lifestyles, manages to implement the new materials developed by researchers, but not yet enhanced into concrete forms of innovation, through methods of application and interpretation that respond to unresolved needs set out in the contemporary context. Design research can guide and modulate the relationships between experimental and industrial dimensions, helping to increase the TRL, the degree of maturity of the materials developed in engineering and chemical fields, in industrial production and favoring the creation of innovative start-ups (Peters, 2014). In this case, designers manage to understand the opportunities as well as the limits of the new materials so as to use them in contexts in which the identity and potentialities can be best expressed and acquire value.

Design is able to build a bridge between the scientific research of materials and production contexts through the prediction, realization, and prototyping of research results in the form of applications and products, which amplify the possibility of persuading companies to implement them in their productions (Luo, 2015; Na et al., 2017). Going even further into the intimacy of the development process, some designers manage to push themselves into an almost equal cooperation by making a substantial contribution to the development processes of new materials, proposing changes, and identifying correlations between the parameters that scientists would have missed. They therefore create grafts of unusual components for researchers, which lead to leaps of innovation that only the intervention from outside of an unusual variable such as design can induce.

In proposing these deviations, designers often consider the needs of the market and the user affordances or try to transfer their experience-driven visions of innovative products by including the development phase of the materials into a single organic path (Barati & Karana, 2019). The design of materials, in these cases, should be based on a mutual cooperation between designers and scientists based on the scientific method, rigorous protocols, as well as interdisciplinary cooperation methods that allow to conceive a composite and multifaceted material experience in which both the technical performances flow and the perceptive, emotional, and evocative qualities that designers care about are involved. The concept of material experience extends to the conception phase of materials, which becomes an experience of collaboration and contamination between designers and scientists, made of comparisons, exchanges, references, divergences, and convergences.

The advantages of a mutual relationship between design and materials science can be found in the possibility for scientists to take advantage of the predictive and interpretative ability of the designer, who is used to moving in areas of application that connect science and society. At the same time, designers can, through the scientific approach, undertake feasible and concrete experimentation and innovation paths, balancing their innate tendency to imagine material solutions that are impossible to achieve or to delegate most of the product innovation to materials without however fully knowing their properties and potential. This constructive and revolutionary form of collaboration can only take place with work opportunities in close contact, in the practice of the scientific laboratory, and exchanging information necessary to integrate the two different approaches and knowledge, thus generating new cultural tools.

Through mutual collaboration, designers and scientists compare and share their different points of view on materials. While chemists and engineers perceive materials in an instrumental way, through the investigation and production devices that are mostly based on two-dimensional images, numerical data, and formulas, designers appreciate materials as the body, sensual, and multisensory components, and therefore as a provider of experiences. The awareness of scientists about the experience dimension of materials is one of the most important revolutions induced by this phenomenon of approach between design and science that will lead to materials of the future that are not only technically more advanced but also much denser from a psychoperceptive and neurocognitive point of view.

The world of materials for designers is very different from how it is observed by scientists and technologists because it constitutes a universe of expressive opportunities through which to express ideas and concretize any relative visions. In the imagination of designers, the maps and technical classifications based on the chemical, physical, and mechanical properties of the different types of materials are superimposed on the interpretations of designers and users who graft references to the cultural, symbolic, experiential, imaginative, iconic, and sensorial aspects which outline what Bachelard called the "poetry of materials." Along with the constellations of the technical properties of materials, the culture of design juxtaposes a "connectomics" that unites materials with objects, design thoughts, intentions, and attitudes of the designers who have interpreted the different materials through those objects. A system of profound traces is engraved in the history of the anthropological relationship between man and material.

The new materials experience generated by the collaboration between design and science is highly enhanced, intense, and profound since the intents of the designers in relation to the gradients and perceptive, evocative, and emotional nuances are translated through the tools of science. They are translated not only through detailed control but also the specific chemical-physical characteristics of the materials that modulate perceptions and sensations. The realization of the materials experience is therefore based on parameters elaborated with instrumental tests, chemical-physical manipulations at the molecular scale, digital simulations, or very precise and exact digital manufacturing. The quantitative and qualitative management of the designing of the materials experience also facilitates collaboration with the neurosciences that can study the neurological feedback linked to the metrics of various parameters through tools such as electroencephalogram, providing useful information for coding the experiences, while decoding user reactions. With this approach, users have the opportunity to live even very unusual and original experiences since qualities such as softness, lightness, layering, grain, and texture are the result of the encounter between different points of view can be calibrated according to a hybrid qualitative and quantitative approach, which leads to completely new and unexpected results.

Through the collaboration between design and science, it is possible to obtain materials that surprise because the experience that characterizes them is different from what would be expected, making it possible to see equally unusual applications. The traditional categories of materials experience are no longer viable. Very lightweight and layered metals used for wearable accessories, transparent woods used for separating walls, soft and perfumed stones to sit on, and elastic ceramics to hold food are just some of the experiential combinations of materials and functions that the mutual collaboration between design and science can produce.

The results of the cooperation between designers and material scientists carried out in laboratories are often unusual and original. These include materials such as *bioplastics from the sea* made with a bio-inspired approach from raw materials of marine origin including mussel valves, algae, and prawn shells, developed by the designers Clarita Caliendo and Francesco Amato, with the chemistry research group coordinated by Mario Malinconico. They were put on display at the itinerant exhibition Italy: *The Beauty of Knowledge* promoted by the science center Città della Scienza and the National Research Council (CNR). The characteristics generally conferred by materials chemists are subverted in these materials from the point of view of the designers: discontinuity instead of continuity, inhomogeneity instead of homogeneity, color nuances and opacity gradients instead of chromatic and optical uniformity, while still ensuring the resistance and adaptability of the structural responses to mechanical stresses. The samples of these materials are shown in Fig. 10.2.

Designers and scientists can conceive the material experience together, from the micro- to the macro-scale as in the FARE LUCE project, which was based on a productive and essential collaboration between material chemists and designers. The project focused on ETE technology (Emulsified Thermoplastics

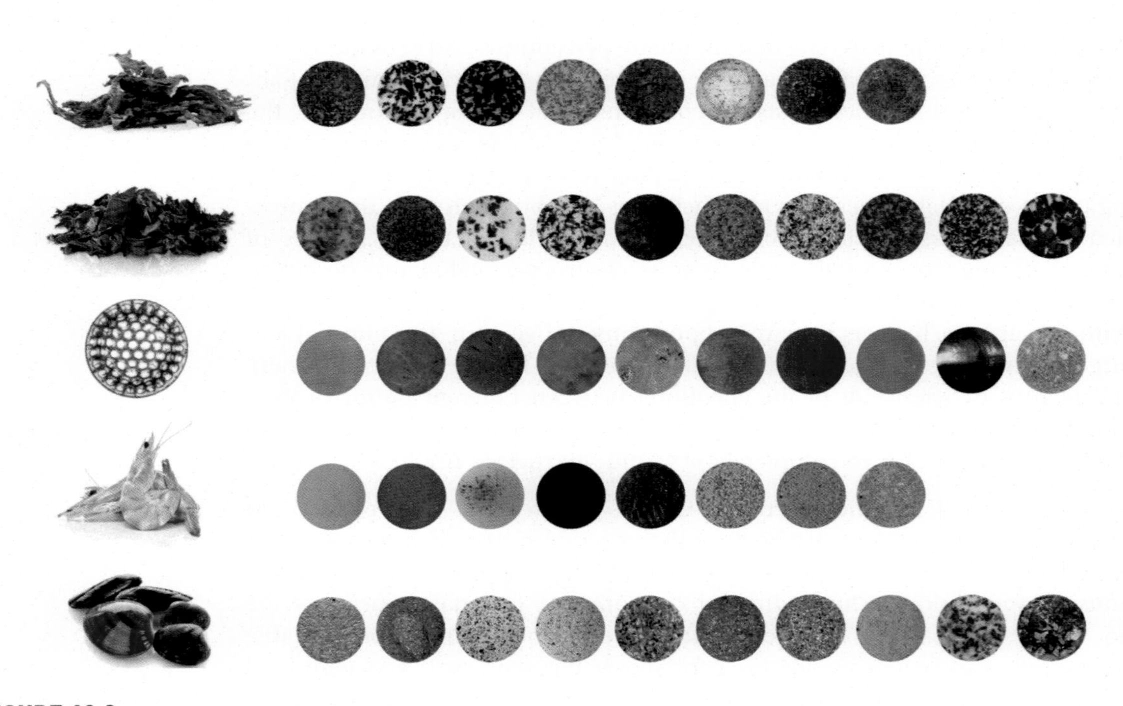

FIGURE 10.2
Material samples of bioplastic from the sea developed by the designers Clarita Caliendo and Francesco Amato, with the chemistry research group coordinated by Mario Malinconico.

Engineering), which was patented by the research Institute of Polymers, Composites and Biomaterials (IPCB) of the CNR with the companies Airpol and Res Nova Die (Aversa et al., 2019; Corvino et al., 2016). The technology uses a polymer (polystyrene, acrylonitrile butadiene styrene, and polycarbonate), sourced from packaging waste or e-waste, loaded with postconsumption waste, in particular, resulting from the grinding of composites, fiberglass, and polystyrene derived from the demolition of multimaterial products such as boats or buildings, to produce a thermoplastic technopolymer through a cold process. The samples of ETE materials developed by the collaboration between designers and chemical scientists are shown in Fig. 10.3.

The designers interpreted this project guiding it in relation to the material experience expected for the final objects. Marzia Micelisopo's intervention in the design of the material developed to make the translucent diffuser panels of the relational lamp called *Eve* from the recycling of buildings led, for example, to making explicit through a metal grain the multimaterial origin to visually communicate the ability of the ETE process to recycle complex

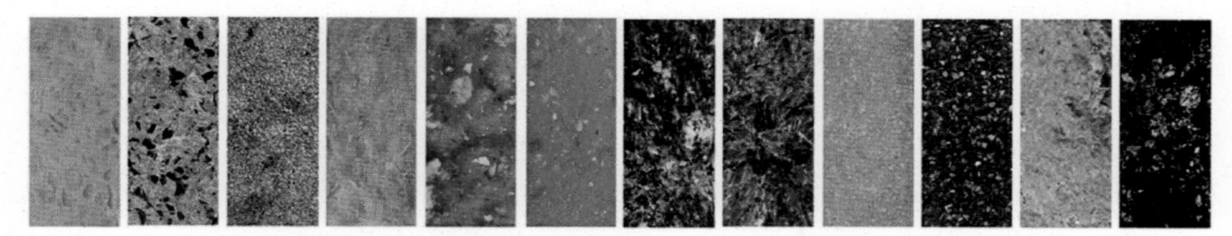

FIGURE 10.3
Material samples developed by the ETE process exhibited in the Airpol—Res Nova Die material library. *ETE*, Emulsified Thermoplastics Engineering.

FIGURE 10.4
EVE, relational lamp designed and realized by the designer Marzia Micelisopo in collaboration with the chemists of the IPCB of CNR.

products without separating materials. The chemists on their own would have mixed the different materials making them unrecognizable or hidden, whereas through a metal the origin of the recycled material is communicated (Fig. 10.4). The designer also proposed an analogy between the regeneration of human relationships and materials. The degree of transparency, opacity, and colors has also been carefully and strategically studied to obtain an effect similar to a hearth evoked by the lamp as illustrated in Fig. 10.5.

In the *Fold* lamp, however, the designer Martina Marchi, the chemists of the CNR and the engineers of Airpol and Res Nova Die, dosed together the layers of material and the way of juxtaposing them to obtain different light intensities in different directions to obtain a directionally differentiated light (Fig. 10.6). The lamp was developed to represent, through the object and the material experience, lights and shadows of the local industrial context and the importance of combination and stratification between production history and innovation.

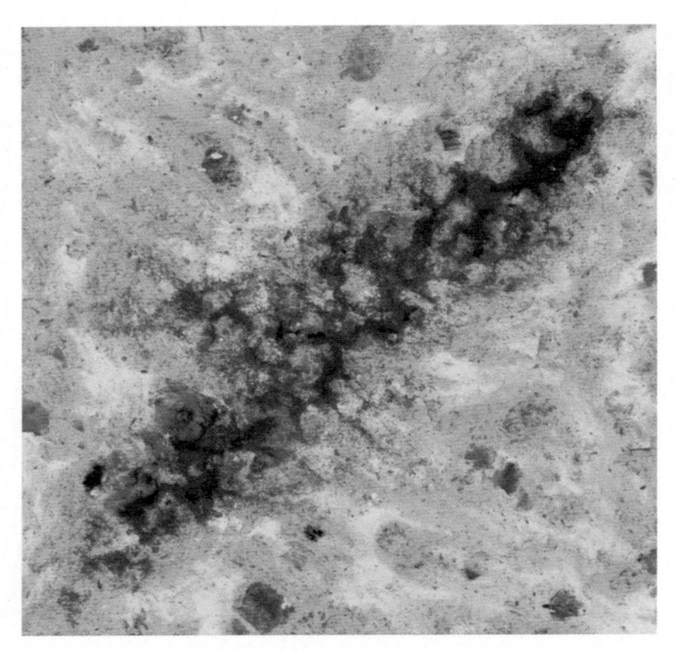

FIGURE 10.5
Detail of the material specifically developed for *EVE* lamp, from the upcycling of building wastes. *Realized by the designer Marzia Micelisopo in collaboration with the chemists of the IPCB and Gabriele Pontillo.*

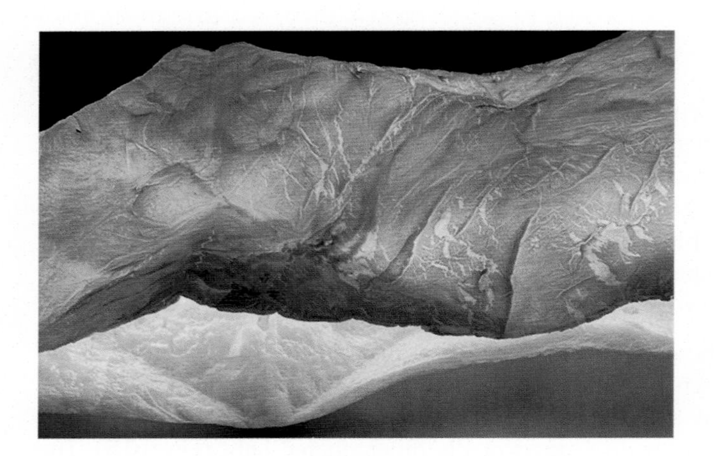

FIGURE 10.6
Detail of the material specifically developed for the *Fold* lamp, made with expanded polystyrene waste structured in overlapping layers. *Realized by Martina Marchi in collaboration with the chemists of the IPCB and the companies Airpol and Res Nova Die.*

10.5 Conclusion

Design that manipulates materials by collaborating with science can enrich the materials experience with a significant expressive, narrative, and experiential potential that is rooted in the instinctive, sensorial, ancestral, and involuntary memory of humans. Through a scientifically competent approach, it acquires the opportunity to explore and define new material relationships, combining the technical and sensorial aspects of the materials, to create richer product experiences, based on the observation and knowledge of biological, chemical, and physical principles (Thompson & Ling, 2014). The contribution of neuroscience to this type of intervention can be very valuable. Many of these topics are studied in the context defined as "embodiment" based on interdisciplinary research that integrates studies on the phenomenology of the living body with neurosciences and cognitive sciences, to understand the complex relationships between body factors, such as physical and motor perception and cognitive processes (Van Rompay & Ludden, 2015).

The digitization, virtualization, and dematerialization of experiences are allowing people to evade physical contact with others and things, with this probably leading, through an opposite reaction, to making people more sensitive and permeable to material experiences. This must provide designers with a greater awareness of having new possibilities available to affect the emotions, the appeal of the unconscious, and the induced thought of users through materials experience. The design of the materials experience can become a very powerful tool if based on the knowledge of chemical and neuropsychological mechanisms, on the study of pheromones and sensory perceptions, which can also prove useful in balancing the evanescence of the digital, through the body component to safeguard the size of physical things and thus to ensure the psychophysical balance of people.

In all these forms of designing materials and material experiences, the visionary nature of design emerges. Through it, unconventional creative strategies manage to facilitate the dissolution of barriers and the building of paths, in which design and science come closer, to fertilize and eventually hybridize, ultimately helping each other out in a coevolution. Design becomes a tool to bring people closer to the progress of science that affects the quality of life, through experiences rather than products, which involve aspects such as curiosity, fun, and emotion, not remaining on the surface, but rather to impress more profoundly the planning thought in the materials of things.

Acknowledgments

The FARE LUCE project multidisciplinary research team involved material chemists: Mario Malinconico, Salvatore Mallardo, and Maurizio Avella of the IPCB of the CNR; designers: Marzia

Micelisopo, Martina Marchi, Alfredo Buccino, Carla Langella, and Gabriele Pontillo; a design historian: Francesca Castanò; an expert of cognitive sciences: Maria D'Ambrosio; and experts of lighting technology: Sergio Sibilio and Roberta Laffi. The involved companies are Airpol and Res Nova Die.

References

Agapakis, C. M. (2013). Designing synthetic biology. *ACS Synthetic Biology, 3*(3), 121–128.

Anderson, C. (2012). *Makers: The new industrial revolution*. New York: Crown Business.

Antonelli, P. (2008). *Design and the elastic mind*. New York: The Museum of Modern Art.

Aversa, H., Rognoli, V., & Langella, C. (2019). *Re-designing recovered materials. Case study: Fiberglass in the nautical sector*. In *Proceedings of the 3rd LeNS world distributed conference* (pp. 884–889). Milan: POLI.design.

Barati, B., & Karana, E. (2019). Affordances as materials potential: What design can do for materials development. *International Journal of Design, 13*(3), 105–123.

Bellotti, E., Kronegger, L., & Guadalupi, L. (2016). The evolution of research collaboration within and across disciplines in Italian academia. *Scientometrics, 109*(2), 783–811.

Caliendo, C., Langella, C., Santulli, C., & Bove, A. (2018). Hand orthosis designed and produced in DIY biocomposites from agrowaste. *Design for Health, 2*(2), 211–235.

Corvino, R., Malinconico, M., Errico, M.E., Avella, M. & Cerruti, P. (2016). New life for aircraft waste composites. In *Proceedings of ANTEC—annual technical conference of the Society of Plastics Engineers*, Indianapolis.

Gaukroger, S. (2006). *The emergence of a scientific culture: Science and the shaping of modernity 1210–1685*. Oxford: Oxford University Press.

Habermas, J. (1986). Der philosophische Diskurs der Moderne: Zwolf Vorlesungen. *Suhrkamp*.

Hekkert, P., & Karana, E. (2014). Designing material experience. In E. Karana, O. Pedgley, & V. Rognoli (Eds.), *Materials experience* (pp. 3–13). Oxford: Butterworth-Heinemann.

Ito, J. (2016). Design and science. *Journal of Design Science*. <http://jods.mitpress.mit.edu/pub/designandscience> Accessed 30.06.20.

Karana, E., Hekkert, P., & Kandachar, P. (2008). Materials experience: descriptive categories in material appraisals. In *Proceedings of the conference on tools and methods in competitive engineering* (pp. 399–412). Delft: Delft University of Technology.

Kemp, S. (2017). Design museum futures: Catalysts for education. *Futures, 94*, 59–75.

Kotha, R., George, G., & Srikanth, K. (2013). Bridging the mutual knowledge gap: Coordination and the commercialization of university science. *Academy of Management Journal, 56*(2), 498–524.

Kutz, M. (2003). *Standard handbook of biomedical engineering and design*. New York: McGraw-Hill.

Langella, C. (2019). *Design e scienza*. Trento: ListLab.

Luo, J. (2015). The united innovation process: Integrating science, design, and entrepreneurship as sub-processes. *Design Science, 1*(e2), 1–29.

Na, J. H., Choi, Y., & Harrison, D. (2017). The design innovation spectrum: An overview of design influences on innovation for manufacturing companies. *International Journal of Design, 11*(2), 13–24.

Nietzsche F. (1872). *Die geburt der Tragödie aus dem Geiste der Musik*. Leipzig: E. W. Fritzsch.

Olson, G. B. (2000). Designing a new material world. *Science, 288*(5468), 993–998.

Oxman, N. (2016). Age of entanglement. *Journal of Design and Science*. <https://jods.mitpress.mit.edu/pub/AgeOfEntanglement> Accessed 30.06.20.

Parisi, S., Rognoli, V., & Sonneveld, M. (2017). Material tinkering. An inspirational approach for experiential learning and envisioning in product design education. *The Design Journal, 20* (Suppl. 1), 1167–1184.

Peters, S. (2014). *Material revolution 2: New sustainable and multi-purpose materials for design and architecture*. Berlin: Walter de Gruyter.

Ramsden, J. J., Allen, D. M., Stephenson, D. J., Alcock, J. R., Peggs, G. N., Fuller, G., & Goch, G. (2007). The design and manufacture of biomedical surfaces. *CIRP Annals, 56*(2), 687–711.

Rognoli, V., Bianchini, M., Maffei, S., & Karana, E. (2015). DIY materials. *Materials & Design, 86*, 692–702.

Schifferstein, H. N., & Wastiels, L. (2014). Sensing materials: Exploring the building blocks for experiential design. In E. Karana, O. Pedgley, & V. Rognoli (Eds.), *Materials experience* (pp. 15–26). Oxford: Butterworth-Heinemann.

Seidl, R., Brand, F. S., Stauffacher, M., Krütli, P., Le, Q. B., Spörri, A., Meylan, G., Moser, C., González, M. B., & Scholz, R. W. (2013). Science with society in the Anthropocene. *Ambio, 42*(1), 5–12.

Thompson, R., & Ling, E. N. Y. (2014). The next generation of materials and design. In E. Karana, O. Pedgley, & V. Rognoli (Eds.), *Materials experience* (pp. 199–208). Oxford: Butterworth-Heinemann.

Van Rompay, T., & Ludden, G. (2015). Types of embodiment in design: The embodied foundations of meaning and affect in product design. *International Journal of Design, 9*(1), 9–11.

Verganti, R. (2018). *Overcrowded: Il manifesto di un nuovo modo di guardare all'innovazione*. Milan: Hoepli Editore.

Wilkes, S., Wongsriruksa, S., Howes, P., Gamester, R., Witchel, H., Conreen, M., & Miodownik, M. (2016). Design tools for interdisciplinary translation of material experiences. *Materials & Design, 90*, 1228–1237.

Zuanon, R. (2014). Design-neuroscience: Interactions between the creative and cognitive processes of the brain and design. In *International conference on human-computer interaction* (pp. 167–174). Cham: Springer.

CHAPTER 11

Materialdesign: design with designed materials

Markus Holzbach
IMD Institute for Materialdesign University of Art and Design Offenbach, Offenbach am Main, Germany

11.1 Introduction

Since its beginnings in the industrial age of the 19th century, industrial design has developed into an expanded concept of design that today affects many areas of life. The increasing digitalization of our living environment reinforces the approach of thinking in interdisciplinary processes and seeing different areas of society and knowledge as a unity or at least linking them. This offers great opportunities, especially for the design disciplines. Today, design can no longer be classified as a core discipline only in its classical terrain—as Mateo Kries, Director of the Vitra Design Museum, sees design today "as a cross-sectional discipline between art, science and technology" (Kries, 2013, p. 136; quote translated by the author). Today, more than ever before, design and creation can be understood—in addition to its contribution to industrial value creation—as a cultural task.

Materials are indicators of a strengthening or changing will to design. They act as impulse generators for new forms of design and also as incubators of cultural-historical developments. These are significantly accelerated by the use of new artificial materials and processes. The progressive approach of thinking in interdisciplinary processes and seeing material, construction, and design as a unity represents a connecting factor for modeling for many disciplines in our digitized world. Digitization is predestined to bundle such diverse, but mostly different and heterogeneous processes. Only through their visualization and materialization, our ideas and design intentions become visible and tangible—and it is the materials that transfer our ideas into the real world and increasingly take over the role of the actual object. Through the materialization of our ideas and design concepts, they become tangible. Today, design concepts are fundamentally influenced not only by digital design and production tools but also by their materiality in a blending of digital and real.

Materials Experience 2. DOI: https://doi.org/10.1016/B978-0-12-819244-3.00013-2

Materials thus become carriers of a wide variety of information and enter into a dialog with their environment. They become informative and intuitive. The design *with* material becomes a design *of* material and in the consequent continuation, "Design with Designed Materials" (Holzbach, 2015).

11.2 Design with designed materials

In contrast to classical product and industrial design, in material design the processes and also the basic principles of materials, structures, and systems play a decisive role. The aim is to provoke results with a usually high degree of experimental freedom. It is the material that transfers our ideas into the real world and increasingly take over the role of the actual object. The traditional thought pattern of "material-authentic design" seems to have been overcome by the combination of digitalization and materialization. Speculative design processes link the material with various research and creation methods in order to design and inform the materials before they are themselves integrated as part of the design process in a larger context. Simultaneously, or interacting with the previous more rational and analytical—and mostly technological or scientific—methods, creative and design methods are also introduced into the design-oriented development of materials. In contrast, essential partial aspects of the design process, which up to then had mostly been predetermined, now take place on the material level from the very beginning. Functions of the object to be designed are anticipated and transferred to the material. The question of materiality is asked at the very beginning of the design task and not, as usual, as a selection of existing materials at the end of a traditional design process. But it is less about the traditional choice of materials in the sense of wood, ceramic, metal, etc. Rather, the question of the characteristics, sensitive properties, and nature of the material to be used arises. The selection and description of the material profile already includes and subsumes the hybrid properties of the object to be designed. Design takes place with designed materials—it is "Design with Designed Materials."

Despite digital charging or information, the material can still be experienced—instead of pure digitalization, there is an increased "reanalogization." Even in the digital world, material has not—as predicted—lost its importance. Despite progressive digitization, it remains the digitization of our world—a physical world. Rather, an increasing overlapping of the digital with the real, that is, our living world, can be observed. The digital infiltrates the "real" and the physical and material world surrounding us. From the symbiosis of the material and designed objects with the digital and material-technical charges, hybrid conceptions result, which open up new ways of formal and functional design. The path from static to dynamic and process-oriented properties is thus smoothed.

11.3 Sensitive properties

The recognition of the specific sensitive material qualities—beyond given material parameters such as tensile strength or modulus of elasticity—is the great potential for designers. These are qualities in the material that are difficult to describe theoretically. This logic of the material is interesting for designers. The interpretation of material design is also about the integration of material research. Especially the integration of material parameters in digital models, as it is possible to see them more and more in design disciplines, originates from the natural sciences and technically motivated disciplines. Of central importance are not only functional but also esthetic qualities. Therefore in addition to the material as such, as well as the manufacturing processes and construction, "soft" factors such as emotions or sensitive properties are of decisive importance. What does the texture, the aggregate state, elasticity, or viscosity of a material tell us? (Holzbach et al., 2014).

By placing the materials in completely new contexts or by digitally and procedurally charging and developing hybrid material combinations, the materials are given new information. Today, generative methods can be seamlessly integrated. In the design process, digital design and manufacturing tools can be used in continuous process chains, not least because of the appropriate scaling. Nevertheless, there is still the physical handling of the material, the constant questioning of the concept as well as the matching and interaction between digital and analog. Historical production techniques are combined with new digital methods, or even lost knowledge is transferred with new contemporary digital tools into a new context and thus also newly informed.

The essence of the material itself becomes increasingly—and then also often intensified—sensually perceptible again. Otl Aicher spoke in 1987 in his essay "greifen und begreifen" about the grasping and understanding of things (Aicher in Braun, 2005). The reinterpreted material worlds open up a variety of ways to shape our environment. In addition to the principle of the minimal, the experimental methods for finding forms and the observation of natural and technical structures, the self-organizing processes and the finding of forms, the focus is on actively shaping the processes taking place. These seem to contradict each other, as Frei Otto already described in 1972: "The will to emphasize design is in contrast to the search for the still unknown form which is subject to the laws of nature" (Otto in Burkhardt, 1984; quote translated by the author). Here there is an important superimposition on the current development, in which an intentional will to design takes place through minimalization, but which is inherent in the materials and the material world as hidden information due to its microscaling.

11.4 Informed materials

In a complex setting of highly different, often simultaneous and sometimes even contradictory levels of action, hybrid design processes are initiated, which in their consequence also lead to hybrid design forms with interactive, connective, smart, or gradually varying properties. The new logic of the material often no longer has anything to do with originally "inherent" or even "authentic" qualities. This fundamentally changes the nature of design concepts. The observed linking of the digital with the real or the digital with the material was followed by hybrid structures with their own hybrid property profiles. This results in a formal and functional charging of the existing materials. Classical categories of material are increasingly dissolving. Thus Sabine Kraft writes in her article "Werkstoffe. Eigenschaften als Variablen": "The relationship between form and material has become as diverse as it is ambiguous. A recourse to clear specifications as to what can be conceived and constructed in which material and how, and what esthetic message would be transported by this, is hardly possible anymore—if it ever existed" (Kraft, 2004, p. 24; quote translated by the author). Schäffner (2016) raises an important related point.

> If this quite different, far more active role of materials that function like automatons and represent their own operative systems is currently emerging in the context of material sciences, then it is something that will change our entire understanding of materiality and design, of technology and nature. Material and matter will no longer be regarded as passive masses that become the carriers of technical or symbolic operations. Rather, it reveals itself as an active operative system. (Schäffner, 2016, p. 29; quote translated by the author)

Through digital or process-related information of materials, their formerly material-specific or "authentic" properties are supplemented, overwritten, or completely replaced. This fundamentally changes the nature of design concepts and leads to completely new links of knowledge. "Informed" or "charged" materials with sensitive or smart properties lead to an increasing blending of different contexts. This results in hybrid forms which no longer allow separation. The increasing microscaling has already been described by Nicholas Negroponte in "Beyond Digital" as computers "... that disappear into things that are first and foremost something else. ... Computers will be an important but invisible part of our everyday life" (Negroponte, 1998; quote translated by the author).

New qualities result solely from the microscaling of the individual components and their hidden present—and operating properties. The contemporary and newer generation of digital and also postdigital material hybrids show the reactive properties of the animated world, but suggest the properties of the "intrinsic." In their 1977 film "Powers of Ten," Charles and Ray Eames in a certain way anticipate the question of scaling and the microscaling which is associated with digitization and postdigitization. At the interface between design and science, they, as designers, take a short journey to present the different dimensions of the most distant and largest, as well as the very near and smallest, structures. As Ray Eames writes "Charles learned from Eero Saarinen how important it is to look at things from the next largest or next smallest scale" (König, 2005, p. 85; quote translated by the author).

11.5 Institute for Materialdesign IMD case studies

The Institute for Materialdesign IMD at the University of Art and Design Offenbach works in an experimental and interdisciplinary dialog on the analog and digital intersection of visualization and materialization. In teaching and research, works that in their core deal with the role of material in the design process are conceived in different scales. Many works are the result of experimental, interdisciplinary, and unconventional processes. The different qualities, possibilities, and even impossibilities of materials and their role in the design process are explored. In addition to the digital and analog processes, the special attraction lies in interdisciplinary, material-unspecific combinations and in the experimental transfer from familiar to unfamiliar contexts. Increasingly systems that are inspired by physics, chemistry, or biology are getting developed. A new nomenclature is emerging—materials or hybrid materials receive their information or determination through the manufacturing and forming process or through their compositional joining with other components. The newly interpreted material worlds hence open up a wide range of possibilities for shaping our environment.

Andreas Reckwitz (2010) writes about the "proceduralization of creativity."

> Since the avant-garde movements [...] a special interest has been in the techniques, the procedures of the creative process. These now essentially appear as those of coincidence management [...]. It is not the subject that appears as the original instance of a production process, but it is the process itself that produces something new in its own dynamic. This production of something new can be promoted by certain techniques [...]. For these creativity techniques, the promotion of chance is central—whether in dealing with the material or in the mental sequence of the association. The point is to allow a momentum of processes in which something new is produced [...]. Creativity is then no longer to be identified with a subjective creative power, but amounts to a promotion of unpredictability in dealing with things and ideas. (Reckwitz, 2010, p. 111; quote translated by the author)

The integration of new digital or even material-technical components creates completely new formal and functional contexts—they are informed or overwritten. The boundaries of design, architecture, art, technology, or science are often no longer recognizable. Many of the works created in this way move at the interface between human and material or nature and artifact. Emerging hybrid forms no longer allow a clear separation or even classification. Partly different and also contradictory materialities and functionalities coexist alongside and with each other. High-tech is linked with low-tech strategies and nonlinear storylines are implemented. Materials increasingly have properties that seem to be "alive." At the same time, natural materials are linked with synthetic materials and digital interfaces and thus transformed into the artificial world. Materials or hybrid composites with sensitive, smart, or gradually varying properties lead to new and complex design concepts. The aim is not a dogmatic reduction but a conceptual materialization based on a variety of analog and digital methods and their combination. New "highly charged" materials increasingly take over the role of the actual object. The observable linking of the digital with the real or the digital with the material leads directly to hybrid structures with their own hybrid property profiles. The following case studies were developed in research and teaching at the Institute for Materialdesign IMD at the University of Art and Design Offenbach.

11.5.1 Parametric Skin

Parametric Skin is a structure that enables us to experience leather in a new way. The design superimposes a graphic, computer-generated honeycomb structure on the natural microstructure of leather. This artificial structural pattern changes the natural appearance of leather by exaggerating it, in particular, in the transition from two- to three-dimensional. The digital net structure that Parametric Skin is based on is made by parametric programming of the organic leather grain. This is why it adapts to any surface shape. The distortion of the structure emphasizes the edges and specific areas of an object. Using several such "informed" leather areas, complex spatial objects without seams can be made. The leather gives objects such as the inherent stability though they remain flexible (Fig. 11.1).

11.5.2 Magnetic Fabric

The work Magnetic Fabric explores the properties of a textile such as mobility and flexibility. Integrated, magnetically effective components are used according to various parameters and depending on the arrangement of the elements and the nature of the textile and set the textile surface in motion. The interaction of active and passive units, methodically arranged inside the textile envelope structure and the surface, causes a mechanical attachment of the components and thus a dynamic rearrangement of the entire medium. The behavior of the

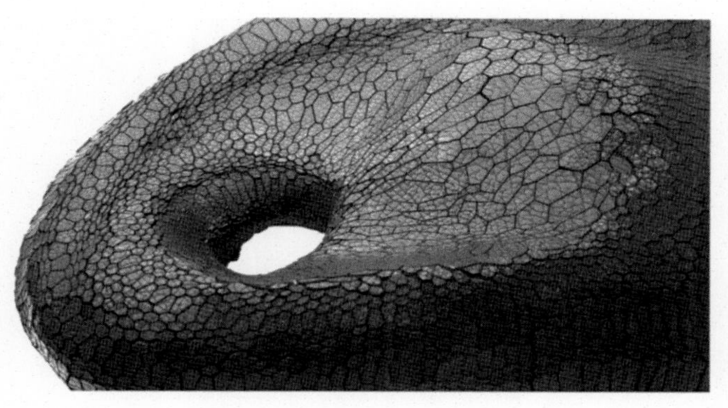

FIGURE 11.1
Parametric Skin. Johannes Wöhrlin at Institute for Materialdesign IMD, University of Art and Design Offenbach. IMD in collaboration with BASF designfabrik and Hyundai Motor Deutschland GmbH.

FIGURE 11.2
Magnetic Fabrics. Lilian Dedio at Institute for Materialdesign IMD, University of Art and Design Offenbach. IMD in collaboration with BMW AG "Intuitive Brain".

textile and the permanent transformation of its shape are initiated by a hidden digital interface, thus suggesting a moving life of the textile itself (Fig. 11.2).

11.5.3 Interactive Wood

Wood as a high-quality and natural material has anisotropic properties and often very individual grains. Each piece of wood is individual and unique. Through touching the surface, the light is activated on-spot, creating a

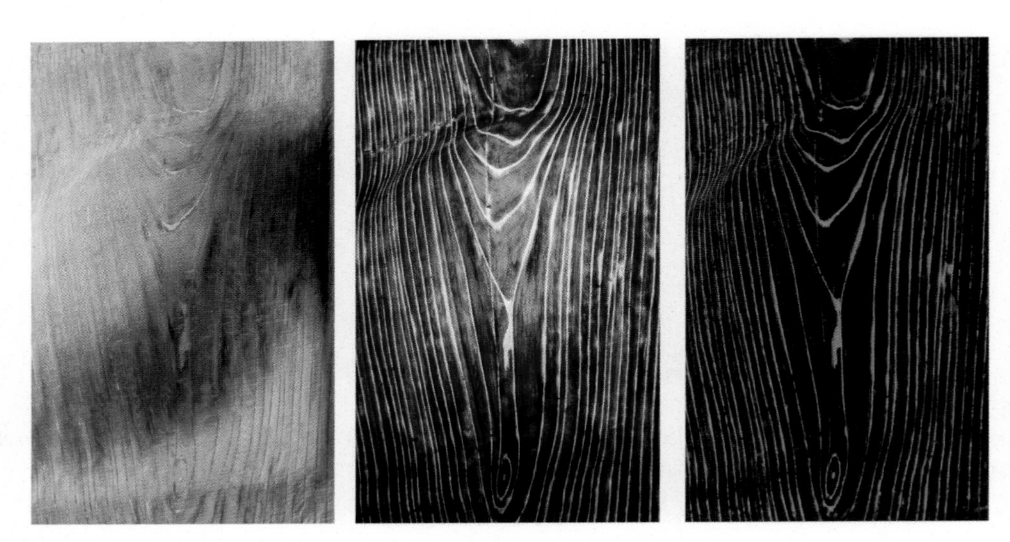

FIGURE 11.3
Interactive Wood stages 1–3. Johannes Wöhrlin at Institute for Materialdesign IMD, University of Art and Design Offenbach. IMD in collaboration with BMW AG "Intuitive Brain".

gesturally controlled functionality. After activation the light is dimmed over time and the grain emits a soft light for a further while. The esthetics of the high-quality wooden surface is visible even in the dark. The wood grain spreads a faint shimmer that offers light for orientation (Fig. 11.3).

11.5.4 Light Skin

Light Skin is the vision of an interactive "skin" in an automotive context. Stimuli such as touch, pressure, and vibrations are received via embedded silicone lenses and released as spots of light: touch becomes visible and provides an optical feedback. The display becomes an analogous interface between object and space. Along with its tactile qualities, in an automotive context Light Skin makes it possible to control the car body and use it as an information and warning display in traffic by visualizing physical forces. Centrifugal and inertial forces would then make the vehicle light up in different ways. Light Skin strives to overcome the existing notion of an automobile and reinterpret its function and appearance (Fig. 11.4).

11.5.5 Hydro Lighting Surface

Hydro Lighting Surface stands for a membrane construction with local function allocation. A resin coating, which is applied to a textile by means of the screen-printing process, enables the opposites solid/flexible and hydrophilic/hydrophobic to be inscribed in a textile fabric. The adhesive

FIGURE 11.4
Light Skin. Julian Schwarze, Martin Pohlmann at Institute for Materialdesign IMD, University of Art and Design Offenbach. IMD in collaboration with BMW AG "Intuitive Brain".

property of the textile is changed in such a way that water adheres to the printed structures like pearls. Thus the water acts as a control and triggers further functions, such as hydrochromic or thermochromic effects. In this way, the coating reacts to the chemical composition of the water or its temperature and makes hidden things visible. Fluorescent pigments contained in the liquid give the textile a luminosity in the dark, which is further enhanced by the prismatic effect of the drop of water. The luminous textile is always active when water interacts with the textile. The water drops act as a local conductor of electricity. Where the water can accumulate, the electric circuit closes and an LED starts to light up (Fig. 11.5).

11.5.6 Transformative Paper

Technical or engineering material parameters are often included in digital models. Today, the materials often undergo digital programming themselves. Swelling behavior and the anisotropic nature of machine-made and wood-based paper are often a disadvantage, as they do not guarantee uniform properties in different directions. However, if one takes advantage of this, materials can be completely defined and activated individually depending on their preferred direction. In this example, the paper as starting material was combined with another material to form a hybrid structure that leads to new combined qualities. The adaptive structure reacts automatically and continuously to environmental influences and is fed solely by the "intelligent" behavior of the used starting materials. The postdigital approach—in contrast to the previously presented example of Magnetic Fabrics—no longer has a

FIGURE 11.5
Hydro Lighting Surface. Alix Huschka at Institute for Materialdesign IMD, University of Art and Design Offenbach. IMD in collaboration with BASF designfabrik and Hyundai Motor Deutschland GmbH.

digital interface. The result is a reactive—lifelike—skin that is able to close automatically when it starts to rain. Depending on the humidity of the air, the paper structure transforms into different states—subtly or very clearly. At high humidity, the reactive skin closes completely and begins to glow intrinsically via a further material and not digitally motivated charge. All reactions take place in real time and are reversible. In the case of high dryness, the individual segments return to their original position (Fig. 11.6).

11.5.7 LEM—Intelligent Skin

Mobility between digitality and exploration. Mobility is one of the central achievements and an important feature of our civilization and society. In the age of limited resources, new mobility concepts have become a relevant subject of debate. Especially acquisitions in the era of information technology shape future forms of mobility to pave the way for unprecedented connectivity. The project "LEM—Intelligent Skin" implies eminent potential for the integration of innovative techniques. The surface is a dynamic communication interface between the automobile, its users, and the spatial context. The passive, supportive surface thus develops into an emancipatory, intelligent skin (Fig. 11.7).

11.5.8 plant b

The project "plant b" deals with alternative manufacturing methods in design. The responsible use of resources ("circularity") is an important topic in material design. Therefore sustainable alternatives for the production of digitally

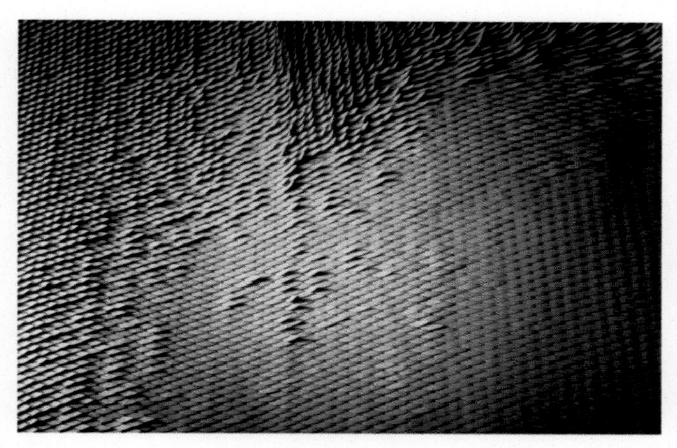

FIGURE 11.6
Transformative Paper. Florian Hundt at Institute for Materialdesign IMD, University of Art and Design Offenbach. IMD in collaboration with BMW AG "Intuitive Brain".

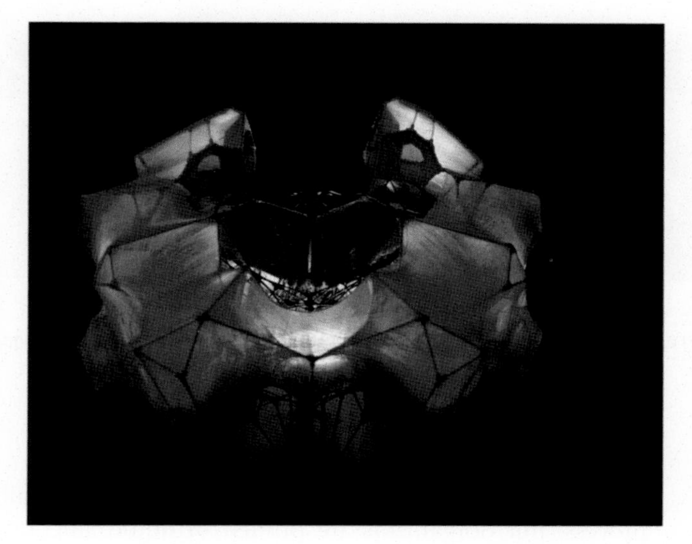

FIGURE 11.7
LEM—Intelligent Skin. IMD: Huschka, A., Grimm, A., Maurer, D., Burfeind, E., Hundt, F., Kovacevic, I., Huisken, J., Wilde, L., Chiera, L., Porstner, L., Mau, M., Riegler, M., Ponce, R, Brück, V., Maskow, V. IMD in collaboration with BASF designfabrik and Hyundai Motor Deutschland GmbH.

designed structures were developed. The 3D-printing process creates complex spatial structures from cellulose that can be combined with growing organisms (wheatgrass and mycelium) and various waxes. The resulting composite is biodegradable and can be returned to the production process (Fig. 11.8).

FIGURE 11.8
plant b. Emilie Burfeind and Andreas Grimm at Institute for Materialdesign IMD, University of Art and Design Offenbach.

11.5.9 MAKU—microorganisms

MAKU is a 3D-printed pneumatic system. In a specially developed printing process, microorganisms are printed in liquid silicone in a special nutrient solution. The symbiotic life cycles of the microorganisms drive the pneumatic system. Different functional and design levels are produced in a continuous printing process. The organisms become more active with increasing temperature and thus drive the pneumatics.

The system functions autonomously and does not require any further aids. Only the ambient temperature must be adapted to the process (Fig. 11.9).

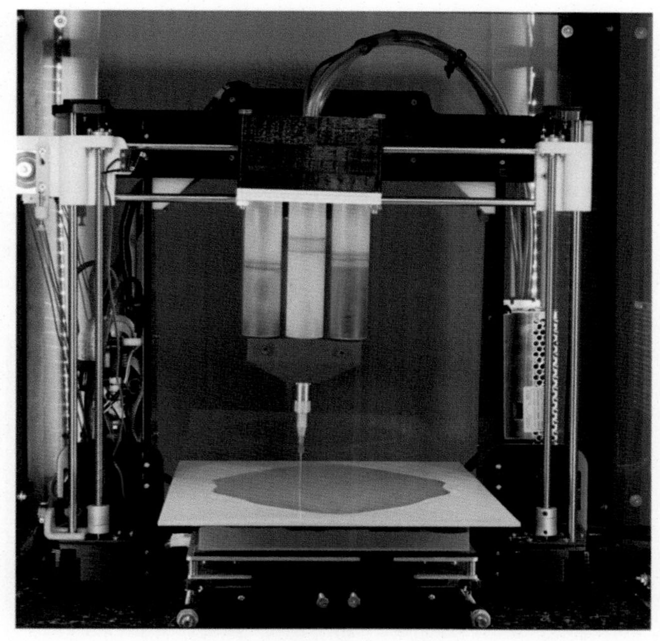

FIGURE 11.9
MAKU Microorganisms. Valentin Brück, diploma thesis at Institute for Materialdesign IMD, University of Art and Design Offenbach.

11.5.10 CeramicWood

Through a thermal process or multistage sintering process, evolutionary wood structures are transformed into biomorphic ceramics. The biomorphic ceramics show the shape of the wood and yet have completely ceramic properties. The transformed wood is cold, very hard, and brittle. Through the transformation process, the warm, rather soft, and ductile characteristic properties of the wood are completely exchanged. Examination with the scanning electron microscope (SEM) shows a ceramic with a completely intact anisotropic wood structure—a hybrid material with a hybrid shape is created. The shape of the wood remains intact at the macro level—likewise, the images taken with the SEM show that the anisotropic structure of the wood remains intact at the microlevel as well. The shape and structural characteristics of one material are combined with the materiality and specific material properties of the other.

The superposition of shape, materiality, and structure from different material areas and the combination of ceramic properties with the special structural conditions of the anisotropic wood structure opens a dialog in which nature with its evolutionary strategies and structures is linked with that of the artificial, perhaps cold efficiency of a material. The resulting image is no longer congruent with the expected and experience-based properties (Figs. 11.10 and 11.11).

11.5.11 Engelstrompete

Presented in the Palmengarten Frankfurt the "Engelstrompete" combines low-tech with high-tech strategies. Natural and artificial components combine to form a light, flat supporting structure made of renewable regional hazel and willow woods and a translucent membrane. The interactive installation interprets the

FIGURE 11.10
CeramicWood Micro. Research at Institute for Materialdesign IMD, University of Art and Design Offenbach, Markus Holzbach/Werner Lorke Mikro_Surface biomorph SiC-Ceramic.

FIGURE 11.11
CeramicWood Macro. Research at Institute for Materialdesign IMD, University of Art and Design Offenbach, Markus Holzbach/Werner Lorke Makro_biomorph SiC-Ceramic.

idea of sustainability: all materials can be easily separated, while no environmentally harmful compounds are used or produced. The pavilion shows analogies to an organism that reacts to its environment and communicates with it—visually and acoustically. The very light translucent membrane forms the skin between the wooden leaf ribs and at the same time serves as a reflection surface or space for different projections. Moving forms and changing colors underline the analogy to a living organism. The visual level interacts with the visual sound space, which was specially designed for the project. Sounds of nature, the twittering of birds from the surroundings, but also urban sounds, for example, from the neighboring underground railway, are the raw material for the acoustic superimposition and are reproduced in an alienated way. The video projection, which is adjusted to the sound situation, interacts with the metallic shimmering inflorescences that are moving in the wind and changing color and shape progressions (Fig. 11.12).

11.6 Hybrid—material, properties, . . ., communication

"Hybrid" material solutions integrate digital and material-specific technologies at a scale and complexity that no longer allows conclusions to be drawn about the performance or function of the existing material systems. Materials become informative or even intuitive, enter into a dialog with their environment and thus possess new functions of meaning. Many of today's object and material hybrids follow the approach of sustainability and focus on an intelligence that is not always inscribed in the material itself, but in the way it is constructed, joined, and used. The digital, in particular, is predestined to

FIGURE 11.12

Engelstrompete. IMD-Project for Luminale Palmengarten Frankfurt am Main IMD: Freund, A.; Kliem, P.; Wildung, B.; Würkner, B.; Bagdulin, N.; König, P.; Lilienthal, S.; Pohlmann, M.; Reinhardt, N.; Sound: Eulberg, Dominik; Foto: Wabitsch, Emily.

connect different sensory levels. In the digitalized world surrounding us, concepts are increasingly developing in the intermediate area between Art and Science or Design and Technology.

Real materials are digitally or procedurally charged, programmed, and informed and increasingly possess dynamic and "smart" properties. How much do nature and material interact with the artificially created? The real and the virtual combine in new hybrid forms with their very own hybrid characteristics. How and with which tools can these be described? Materials are reprogrammed by placing them in completely new contexts, digitally charging, and developing hybrid concepts. Through this information, materials acquire new properties and attributes of "living" or "intelligent." They become interactive, connective, or even smart—they seem to react automatically to their environment. Microscaling thus leads to the suggestion of apparently existing formal and functional properties in the material. The resulting static, dynamic, and process-oriented properties, some of them existent together, give rise to new design concepts and, at the same time, important questions. How do different and sometimes contradictory characteristics act and interact in hybrid overall concepts? This is also the ambivalence of the presented material solutions. Shape and properties are no longer congruent.

Articulation of the new material solutions and hybrids will be an important task of material design and creation (Parisi et al., 2020). In classical product design, the instructions for use are clearly communicated through the specific formal training of the sign function. This understanding is essentially based on the previous experience, but also—and here it becomes more speculative—on expectations. Are these clear assignments of function still given in the newly coded and informed hybrids? The role of the sign function as a mediator has to be renegotiated especially in the area of "intelligent," smart, and connective material hybrids. The basis of the knowledge mediating dialog between human and object or also human and hybrid materials lies in experience and recognition.

The hybrid material systems no longer convey these distinct characteristics clearly and the idea becomes ambiguous. What are the signs of the designed? What does it represent? How can it be used? How can hybrids, which are simultaneously located in the real and virtual world, be described with previous tools? Particularly in completely new functional and technological contexts, specific sign functions are often not recognizable, especially at the beginning. There is just as much potential for new solutions in their redefinition as in their transfer from other contexts. They are not necessarily based on stored experiences and the resulting recognition. This has a great impact on communication and the perception of the world around us, characterized by a communication that is no longer based on the necessary recognition and mediation, but that may not take place at all—intuition takes the place of instruction.

11.7 Nomenclature_Interdependence and In-Between

The resulting works often move at the interface of nature and artifact, analog and digital. What is real or fake, natural or artificial? Material simultaneously takes on another role—the role of the actual object. Materials and objects of the material world surrounding us are more and more informed. Through this new "charging level," materials and things are given a new and unique nomenclature—in other words, a hybrid nomenclature.

Hybrid material systems and objects suggest the properties of intrinsic intelligence or the intuitive. The object or hybrid system decides whether it communicates itself to us. This means that it can no longer be a human–object or human–material relationship but conversely an object–human or material–human relationship. Essential factors for the additional arrival of a dialog, such as recognition, now start from the object. This recognition is conveyed by different connectivities and data acquisition on the side of the object—with important user data, user recognition, etc., in order to gain knowledge by creating a user profile and to form the basis for an allegedly "optimal" communication through that. Many of the informed objects and materials are increasingly complex, contradictory, and hybrid. They unite different worlds of the virtual and the real, the physical and the nonphysical. Just as hybrid materials, objects and materials are characterized by hybrid properties, the tools that describe them individually will also have to carry this hybrid nature within them. The dynamic digitization of our real environment world demands a holistic theoretical approach that takes into account the different pluralities and realms of the virtual and the real with the interdependence areas operating in between. Due to the increasing developments in science and technology and the growing interconnection of different fields of knowledge in the digital world, the number of indeterminate interdependencies is simultaneously growing. It is necessary to develop a nomenclature for the in-between. This is linked to the task of expanding and further developing the creative and artistic experiment and interdisciplinary permeability and deriving from these decidedly inductive and hybrid theories. How can processes be described in which microorganisms become active? Bruno Latour writes about nonhuman actors and in this context refutes the subject/object dichotomy. Latour describes this using Pasteur's nutrium solution for yeast cultures as an example:

> Do these belong to subjectivity or to objectivity or to both? None of the above, obviously, and yet each of these little mediations is indispensable for the emergence of the independent actor that is nevertheless the result of the scientists' work. (Latour, 2000, p. 148)

11.8 Outlook

The opening up to other fields of knowledge and science defines a new scientific claim, according to which design is also to be understood essentially as a cultural task and as a cross-sectional discipline in interdisciplinary exchange that has an active role as an impulse generator and designer of our environment. Despite all the desire for determination, the degree of variables, discrepancies, and transitions is growing—the world is becoming more complex and indeterminate. These uncertainties, however, are also a source of great inspiration and ultimately of great essence for many disciplines and also for design.

Digitization is forcing a stronger interaction with different fields of knowledge. This represents a great opportunity, especially for design as a complex image of an even more complex living world. The simulability of intuitive and systemic design processes resembles a multitude of different vectors in space, which take on a variety of—even contradictory—directions. It is important to understand this as an essential quality of a discipline and to allow for individual, independent, or more autonomous interpretations as part of a whole. Design can very well continue to act in the sense of a core discipline and/or a cross-sectional discipline. Linear, solution-oriented, even efficiency-determined, clear industrial design is just as possible here as non-linear studies and experiments—or even contradictory hybrids of these different systems. Intuitive and systemic design processes can take place equally, sometimes autonomously, sometimes interacting. Regular and linear systems that integrate chance or non-linear and disruptive moments and confront the existing system with the question of cause and effect.

From the manifold links and tasks, an increasingly heterogeneous overall profile of the design discipline is derived, based on simultaneity. Design as a discipline that sees itself as the designer of our environment, which, in addition to fulfilling its function and the interest in results, is open to processes, experiments and innovations, nonlinearities, imponderables, contradictions, coincidence, and disruptive moments as well. In the spirit of Ludwig Wittgenstein, this is where new questions arise. Essential aspects concern the questions of material and energy consumption, the digital and procedural charging and information of materiality and things, or digital and postdigital interactive materials, lightweight construction, as well as natural constructions or building with or according to nature and sustainability. Last but not least, all this leads to the central questions of how the human beings will live in the future or what forms of mobility they will have. Thinking in terms of natural processes is very topical. In times of dwindling resources, the minimization of material and energy consumption is more urgent than ever.

References

Aicher, O. (1987). greifen und begreifen. In J. W. Braun (Ed.), Die sprache der hände (pp. 40–41). Brakel: FSB Franz Schneider Brakel. (2005).

Holzbach, M. (2015). Gestalten mit gestalteten materialien. Bauwelt, Ausgabe, *20*, (pp. 26–27).

Holzbach, M., Kracke, B., & Bertsch, G. (2014). Material Grove - von traditionellen Materialien zu zukunftsorientierten Materialentwicklungen (From traditional materials to future-oriented material developments). Offenbach am Main: HfG Offenbach.

König, G. (2005). *Eames*. Köln: Taschen.

Kraft, S. (2004). Werkstoffe. Eigenschaften als Variablen. *Arch +*, *172*, 24–28.

Kries, M. (2013). Diskurs: Zukunftswerkstatt Designmuseum. *Domus 1*, 136–137. (first German edition).

Latour, B. (2000). *Pandora's hope: Essays on the reality of science studies*. Cambridge: Harvard University Press.

Negroponte, N. (1998, December 6/12). *Beyond digital*. Wired. <https://www.wired.com/1998/12/negroponte-55> Accessed 30.09.20.

Otto, F. (1972). Das Zeltdach. Subjektive Anmerkungen zum Olympiadach. In B. Burkhardt (Ed.), Frei Otto. Schriften und Reden 1951–1983. Braunschweig/Wiesbaden: Vieweg & Sohn. (1984).

Parisi, S., Holzbach, M., & Rognoli, V. (2020). The hybrid dimension of material design: Two case studies of a do-it-yourself approach for the development of interactive, connected, and smart materials. In T. Ahram et al. (Eds.), *Proceedings of IHSI 2020 Conference* (pp. 916–921), Modena IT. T: IHSI 2020, Vol. 1131. Cham: Springer-Verlag.

Reckwitz, A. (2010). Vom künstlermythos zur Normalisierung kreativer Prozesse: der Beitrag des Kunstfeldes zur Genese des Kreativsubjekts. In C. Menke, & J. Rebentisch (Eds.), Kreation und Depression: Freiheit im gegenwärtigen kapitalismus (pp. 98–117). Berlin: Kulturverlag Kadmos.

Schäffner, W. (2016). Immaterialität der Materialien. In N. Doll, H. Bredekamp, & W. Schäffner (Eds.), + Ultra. Gestaltung schafft wissen (pp. 27–36). Leipzig: E.A. Seemann.

Index

Note: Page numbers followed by "*f*" and "*t*" refer to figures and tables, respectively.